ROMANCING NEVADA'S PAST

Romancing Nevada's Past

Ghost Towns and Historic Sites of
Eureka, Lander, and White Pine Counties

SHAWN R. HALL

University of Nevada Press *Reno & Las Vegas*

University of Nevada Press, Reno, Nevada 89557 USA

Manufactured in the United States of America
Design by Kaelin Chappell

Library of Congress Cataloging-in-Publication Data
Hall, Shawn, 1960–
Romancing Nevada's past : ghost towns and historic sites of Eureka, Lander, and White Pine counties / Shawn R. Hall.
p. cm.
Includes bibliographical references (p.) and index.
ISBN: 978-0-87417-228-7 (pbk.: alk. paper)
1. Ghost towns—Nevada—Eureka County—Guidebooks. 2. Ghost towns—Nevada—Lander County—Guidebooks. 3. Ghost towns—Nevada—White Pine County—Guidebooks. 4. Historic sites—Nevada—Eureka County—Guidebooks. 5. Historic sites—Nevada—Lander County—Guidebooks. 6. Historic sites—Nevada—White Pine County—Guidebooks. 7. Eureka County (Nev.)—Guidebooks. 8. Lander County (Nev.)—Guidebooks. 9. White Pine County (Nev.)—Guidebooks. I. Title.
F847.E8H34 1993
917.93'150433—dc20 93-22777
CIP

The paper used in this book meets the requirements of American National Standard for Information Sciences—Permanence of Paper for Printed Library Materials, ANSI/NISO Z39.48-1992 (R2002). Binding materials were selected for strength and durability.

Frontispiece: Prospect, 1907. (Denver Public Library)

This book has been reproduced as a digital reprint.

This history of Eureka, Lander, and White Pine counties is happily dedicated to my most ardent supporter, my mother

LORRAINE LOREY HALL

She has taken the time not only to be a parent but also to be a great friend—something that is not easy to do, especially with a free-minded and fiercely independent person like myself. This book also means a lot to both of us, as it was our first coadventure in the wilds of Nevada. Thank you, Mom!

CONTENTS

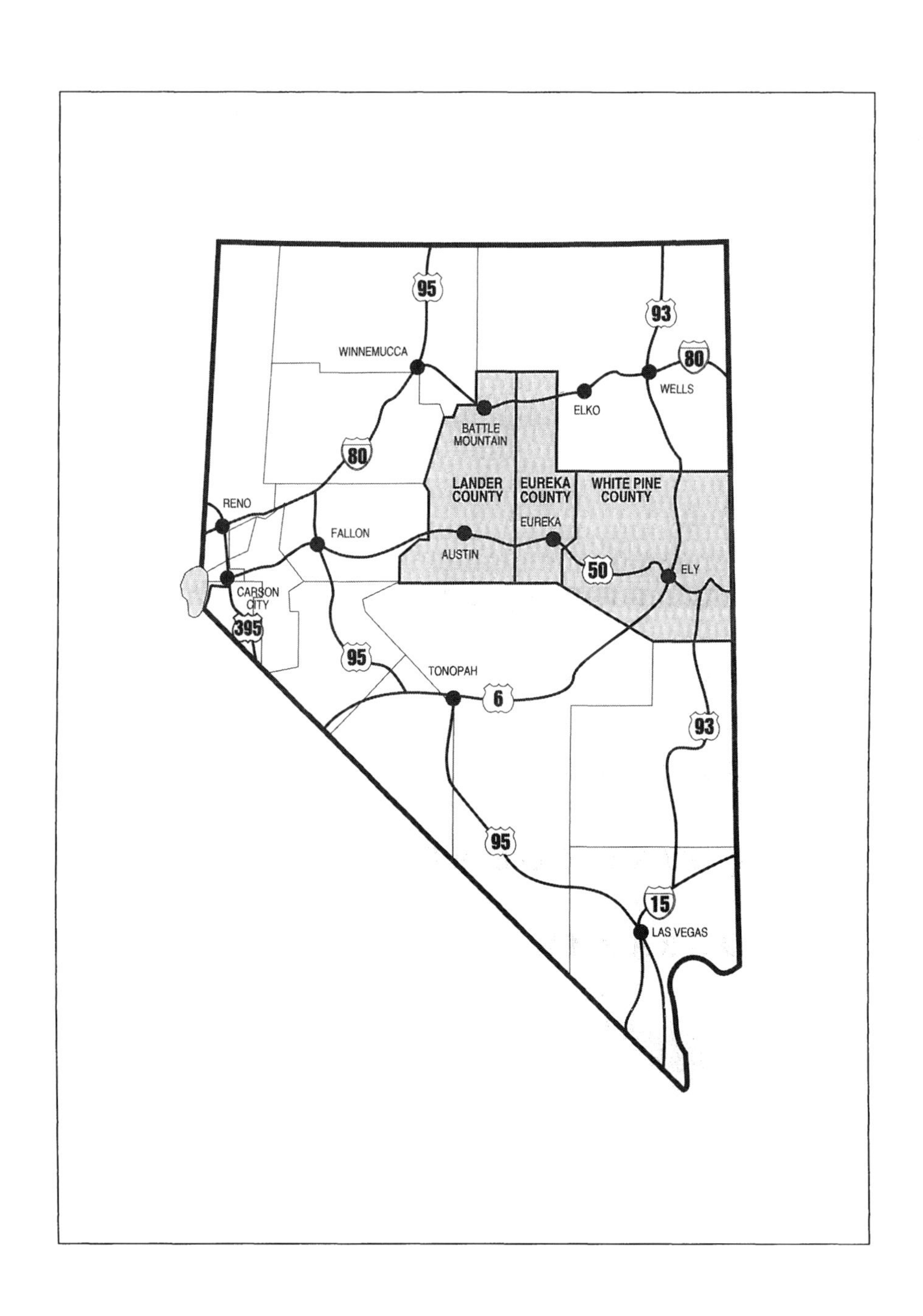
95
93
80
WINNEMUCCA
WELLS
ELKO
BATTLE
MOUNTAIN
80
LANDER
COUNTY
EUREKA
COUNTY
WHITE PINE
COUNTY
RENO
FALLON
EUREKA
AUSTIN
50
ELY
CARSON
CITY
395
95
TONOPAH
6
93
95
15
LAS VEGAS

PREFACE

Although the research for this second volume of my Nevada ghost town series was completed several years ago, finishing undergraduate and graduate studies forced me to curtail my writing activities for a while—and thus the long delay in the appearance of the present study. Many people have asked me about my sequel to the Nye County book, and I am very happy now to have my history of Eureka, Lander, and White Pine counties published. I want to thank all of my earlier readers for their kind letters, and I invite the readers of this book to write to me with any comments or suggestions.

I have striven to cover all of the historical sites in these three counties and to give each place the attention it deserves. Directions to each site are included so that the reader on vacation can visit the wonderful ghost towns that these counties offer. I have included photos with the text and have tried to provide a contrast between the ghost towns at the time of their maximum population and their present condition. While the history of these counties is extremely rich, the scenery offered is spectacular. That alone makes a visit to this area worthwhile, for the beautiful surroundings accentuate the unique appeal of the ghost town settings.

Since many of the historical sites in the three counties are on private property, visitors must take care to respect warning signs. If there is any doubt, one should seek permission from the owner. Please do not tear buildings or ruins apart for relics, or vandalize the old remains. Such senseless acts benefit no one and deprive many others. Take as many photos as you want, but please leave what still stands alone so future generations can enjoy these sites for years to come.

I have already begun work on an Elko County book and hope to have that project completed by 1994. I look forward to meeting many more fascinating Nevadans as I continue my travels through the Great State of Nevada.

Happy Ghost Town Hunting!
Shawn Hall
Elko, Nevada 1993

ACKNOWLEDGMENTS

This book represents a final product that many people have been instrumental in helping to complete.

First, I would like to thank the residents of Eureka, Lander, and White Pine counties, who made me feel very welcome during my travels. This book is a tribute to them and to their ancestors. In particular, I would like to thank Charlie and Lois Chapin, Al and Jean Legarza, Bud Easton, John and Elizabeth Wear, Bob Hanks, and Marie-Teresa Fowell (White Pine Museum).

Photographic assistance came from many quarters: Bob Nylen and Gloria Harjes of the Nevada State Museum, Eslie Cann and Judith Rippetoe of the Nevada Historical Society, Heidi LaPoint of the Western History Research Center at the University of Wyoming, Mati Stephens of the MacKay School of Mines, the Bancroft Library, the Denver Public Library, the United States Geological Survey, the Library of Congress, and the Northeastern Nevada Museum.

Special thanks go to my best friend, Bruce Franchini. As my friend and fellow camping enthusiast, he has joined me in exploring the lost sights of Nevada.

Finally, I want to thank everyone for receiving my first book so well. This book is for you. I hope you will enjoy it.

ROMANCING NEVADA'S PAST

A Short History of Eureka County

Eureka County was created from Lander County on March 1, 1873. Long before any town was established, emigrants were traveling through Eureka County on their way to California. Many journals relate tough times getting across Nevada. Places such as Gravelly Ford were important rest stops for emigrants preparing to cross the dreaded desert.

The core of Eureka County's history revolves around Eureka, the county seat. While initial mining discoveries were made in 1864, it wasn't until 1870 that Eureka began its impressive production. Eureka's prosperity led to a flurry of prospecting throughout the county. Towns such as Mineral Hill, Ruby Hill, and Vanderbilt sprang up. Palisade became an important shipping point for virtually all supplies heading to the south, first for freighters and later for the Eureka and Palisade Railroad.

The Pony Express has an important page in Eureka County's history. Four original stations were located in the county. Freighting and stage lines were prevalent for many years. Eureka and Palisade became centers of goods for towns, mining camps, and ranches in a wide area. The arrival of the Eureka and Palisade Railroad (later the Eureka-Nevada Railroad) in 1875 allowed for the county's continued growth.

Eureka always remained the heartbeat of the county. When Eureka was successful, the county prospered. However, when the major mines of Eureka closed, the county experienced depression-like times.

When Eureka faded in the 1890s, the county fell upon quiet times. Ranching became the mainstay of the economy. While some new strikes were occasionally made—most notably Buckhorn—it wasn't until microscopic gold was discovered in the 1960s that the county was once again at the forefront of Nevada mining production. The incredible amounts being produced from the Carlin Trend in northern Eureka County easily outdistance production numbers from the early years. In addition, new mines in operation near Crescent Valley and Ruby Hill promise to add even more. Eureka County has found a new prosperity that should continue for many years.

Eureka County

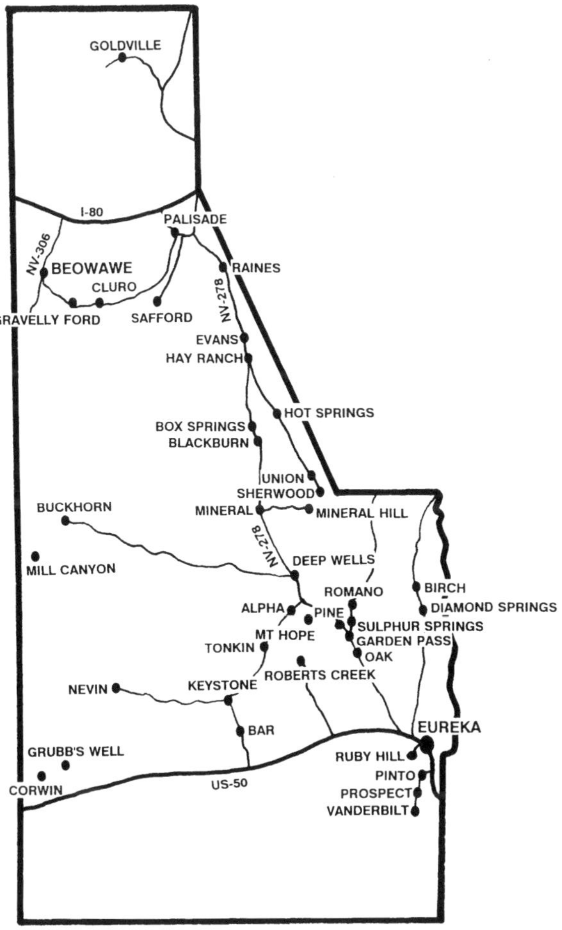

GOLDVILLE
I-80
PALISADE
NV-306
BEOWAWE
RAINES
CLURO
NV-278
GRAVELLY FORD
SAFFORD
EVANS
HAY RANCH
HOT SPRINGS
BOX SPRINGS
BLACKBURN
UNION
SHERWOOD
BUCKHORN
MINERAL
MINERAL HILL
NV-278
MILL CANYON
DEEP WELLS
ROMANO
BIRCH
ALPHA
PINE
DIAMOND SPRINGS
SULPHUR SPRINGS
MT HOPE
GARDEN PASS
TONKIN
OAK
ROBERTS CREEK
NEVIN
KEYSTONE
EUREKA
BAR
GRUBB'S WELL
RUBY HILL
PINTO
CORWIN
US-50
PROSPECT
VANDERBILT

Alpha

DIRECTIONS: *Located 38 miles north of Eureka, via Nevada 278.*

Alpha, the most prominent of the early Eureka and Palisade Railroad stations, flourished because circumstances were perfect for its success. The railroad had proceeded southward from Palisade, but construction was halted at Alpha during the fall of 1874. The boom at Eureka brought hordes of travelers, which allowed Alpha also to boom. During 1874 and 1875, the town was expanded to include a 75-guest hotel, saloons and stores, and all of the Eureka and Palisade Railroad shops and buildings.

W. L. "Nick of the Woods" Pritchard took advantage of this terminus situation and platted a large townsite, gambling that Alpha would remain the southern terminus of the railroad. Pritchard expanded his already huge freighting empire by having his many stages start their southern journey to Eureka at Alpha. Eventually he monopolized the Eureka and Hamilton trade. At his peak, Pritchard had more than 500 wagons traveling the road between Alpha and Eureka.

This ideal situation ended when the Eureka and Palisade Railroad was extended to Eureka during the fall of 1875, ruining any chance that Alpha had for permanency. All the railroad shops were moved to Palisade. Other small businesses moved to nearby Mineral, which was serving the new boom camp of Mineral Hill. The post office, opened on March 20, 1877, closed its doors in February 1886. A nearby ranch used the Alpha name for its post office during the early 1920s, but the town of Alpha had long ceased to exist. Now only faint rubble and sunken cellars mark the lonely site.

Eureka-Nevada engine taking on water at Alpha. (Northeastern Nevada Museum)

Overview of Beowawe in 1922. (Northeastern Nevada Museum)

Beowawe

DIRECTIONS: *Located 6 miles south of Interstate 80.*

Beowawe was an active place long before a town formed. The site was a Northern Paiute campsite for generations. Later, emigrants passed through here during the 1840s and 1850s on their way to California. A small camp formed here during the 1850s around a trading post. The arrival of the Central Pacific Railroad in 1868 led to the actual formation of the town of Beowawe.

Most of the first residents were the Chinese construction workers who were building the railroad. The name "Beowawe" is Northern Paiute for "gate" or "gateway." The town is located at the mouth of a canyon, with an entrance surrounded by high mountains, forming a gateway to the west. There is another version of Beowawe's naming, rooted in legend and not fact. J. A. Fillmore, an early manager of the Central Pacific, was in charge of naming new railroad towns. He was a huge man, who weighed more than three hundred pounds. Supposedly, when local Indians saw him, they ran away yelling, "Beowawe," which nonlocals translated to mean "great posterior."

A post office opened on April 15, 1870. At the time, Beowawe was located in Lander County. Eureka County was created from Lander and the post office transferred to Eureka County on June 18, 1874. That post office is still active today. During the 1870s, Beowawe achieved some prominence as an ore shipping point for the Cortez Mining District, located to the south. While the railroad helped establish Beowawe, ranching became the backbone of the local economy.

Many ranches formed in the area surrounding Beowawe. One of the most prominent is the Horseshoe Ranch. George Grayson, an investor in the Comstock mines, and Andrew Benson formed the ranch in 1872. Benson had come to Beowawe in the 1860s and ran the Beowawe Hotel for many years. Over the years, the ranch continued to grow under the ownership of many different people. From 1936 to 1953, Dean Witter, the prominent San Francisco financier and stockbroker, owned the Horseshoe Ranch. He was always willing to help local people out, and even after he left and became world famous, he never forgot his friends in the Beowawe area. The ranch continues to be successful today.

By 1881 Beowawe had a population of 60 and supported a store and hotel. Through the years, Beowawe's population has remained about the same, rarely reaching more than 200.

After the turn of the century, mining revived in districts to the south. Lou Wilkerson ran a freighting business to these camps and hauled ore from mines in Crescent Valley to the railroad in Beowawe. After the revivals collapsed, Wilkerson and his family moved to Midas, in Elko County.

The boom at Buckhorn, which began in 1909, had a direct impact on Beowawe. While most supplies were shipped to Buckhorn via Palisade, an 800-horsepower power plant was constructed at Beowawe in 1914. Power lines were strung to Buckhorn, about thirty miles south, to power the town and run Buckhorn's 300-ton cyanide mill. However, the boom died in 1916, and the power plant fell into disuse and was finally dismantled.

Today, Beowawe continues to survive and maintains a sleepy existence. The Southern and the Western Pacific railroads' trains still rumble through town, but they don't stop anymore. The old depot was torn down in the late 1980s. Many buildings from Beowawe's early days remain, including the schoolhouse. The huge foundations of the power plant, honeycombed with large tunnels, are very interesting to investigate.

Beowawe boasts a beautiful natural attraction: the Beowawe geysers, located a couple of miles south of town, which have fascinated visitors since emigrant days. With interest growing in the geothermal field, Magma Power Company, Vulcan Thermal Power Company, and Sierra Pacific Power Company heavily explored and drilled the geysers from 1959 to 1965. However, the drilling, with some holes 5,000 feet deep, damaged the natural flow of the geysers. While the flow is less vigorous today, the geysers are definitely worth the trip while visiting Beowawe.

The Buckhorn Mines power plant at Beowawe, 1914. (Mining and Scientific Press)

Birch

DIRECTIONS: From Cold Creek, continue north on Nevada 228 for 13 miles to Birch.

Birch was the site of a small stagecoach station on the Elko-Eureka and Elko-Hamilton stage runs. The station was named after James E. Birch, who had been a pioneer stage driver in northeastern and central Nevada. Even after the stage station had outlived its usefulness, Birch lived on as a stopping place for automobile travelers heading south. A post office opened at Birch on August 20, 1901, and was active until July 31, 1926. Once Interstate 80 was completed, there were paved roads going from Elko to Eureka, and the unpaved road going past Birch received less and less traffic. Today only a couple of collapsed buildings distinguish the Birch site from the surrounding desert.

Blackburn

DIRECTIONS: From Alpha, continue north on Nevada 278 for 16 miles to Blackburn.

Blackburn came into existence in the summer of 1874 as a stop on the Eureka and Palisade Railroad and later the Eureka-Nevada Railroad. The stop was hardly more than a siding until 1909, when the short-lived Buckhorn boom began. Blackburn became the major shipping point for all supplies headed to Buckhorn, and soon the railroad siding began its own boom.

Eureka-Nevada depot and store at Blackburn. Once the Buckhorn boom faded, Blackburn's usefulness was gone. (Northeastern Nevada Museum)

A hotel, a saloon, and a livery stable sprang up to accommodate the rush to Buckhorn. A regular stage line from Blackburn to Buckhorn was started. Unfortunately for Blackburn, the Buckhorn boom collapsed in 1915–16. Blackburn soon faded to obscurity. The only activity that took place at Blackburn after the collapse was a more modern one: A gas station and grill were run here for many years and only recently closed down.

Box Springs (Lodi)

DIRECTIONS: *From Blackburn, continue north on Nevada 278 for 4.5 miles to Box Springs.*

Box Springs was another of the many small railroad stops on the Eureka and Palisade and the Eureka-Nevada railroads. The small siding was formed in summer 1874 and did enjoy a little more popularity than some other sidings. This was the stop for people heading for the Union and Diamond districts to the southeast. A small camp grew up around the tracks, and a fairly steady population of 25 lived here. After the mid-1880s, Union faded and so did Box Springs. By the 1890s, it was once again nothing more than a siding, and served in that capacity until the rails were torn up in 1938. Today nothing at all remains to mark the site.

Buckhorn

DIRECTIONS: From Alpha, head north on Nevada 278 for 5.5 miles. Exit left and follow for 8.5 miles. Bear right and continue for 8 miles to Buckhorn.

The first gold discoveries in what was soon to be Buckhorn took place in the winter of 1908. The discoverers were a band of five prospectors: Joseph Lynn, W. S. McCrea, William Ebert, James Dugan, and John Swan. These men did limited work on their claims. In April 1910, all their holdings were sold to Nevada mining magnate George Wingfield for $90,000. Wingfield immediately formed the Buckhorn Mines Company, with himself as president and E. R. Richards as mine manager.

During the year after the initial discoveries, a small camp formed. On February 18, 1910, a post office opened at the camp, with Laura Wilson as postmistress. By the beginning of 1914, almost 300 people were living in the town, and hopes were high for the future of little Buckhorn. A 400-ton cyanide mill had been completed in 1913, and operations were begun in January 1914. The miners were kept busy trying to fill the large cyanide vats. A 33,000-volt power line was strung to the company's 800-horsepower power plant in Beowawe, supplying both the mill and the townspeople with electric power. The prospects for permanency looked good for the booming town.

The Buckhorn Mill under construction. (American Heritage Center, University of Wyoming)

Alas, not many ore veins are endless, and the Buckhorn veins were shorter than most. Ore production dropped drastically in late 1914 and even more in 1915. The mill was kept running until all hope was gone, but finally, in February 1917, it closed for good. Wingfield's company folded soon after, and the mill was dismantled during the summer. Buckhorn's boom had gone bust.

Buckhorn remained empty, except for an occasional prospector, until 1935. Then several mines were reopened by the Pardners Mines Corporation, with John Banagwanath as president, and the company was renamed the Buckhorn Mining Company. In 1936, an 80-ton flotation mill was constructed. The mill operated from November 1936 until December 1937. This short-lived revival brought only about 50 people back to Buckhorn. By the end of the renewed activity, underground workings put in place from 1910 to 1937 totaled almost four miles. Unfortunately, not much good ore was found despite the tunneling. Total production for the two companies most active in the district is listed at $790,000. Other production by leaseholders and prospectors brought that total to just over $1 million.

Mill foundations and crumbled buildings are scattered throughout the townsite and help to make a visit to Buckhorn worth the long trip.

Cedar

DIRECTIONS: From Oak, continue north on Nevada 278 for 3.5 miles to Cedar.

Cedar was a siding on the Eureka and Palisade and the Eureka-Nevada railroads. The railroad used this site from 1874 to 1938 as a water stop and not much else. No settlement ever formed at the site, and nothing remains at Cedar today.

Cluro

DIRECTIONS: From Beowawe, take good road east for 8 miles. Exit left and follow for 1 mile to Cluro.

Cluro was a stop on the Central Pacific and, later, the Western Pacific, railroads. A small settlement formed here during the late 1870s and early 1880s. A population of 37 lived here in 1881. A few residents continued to live here through the 1920s, but no one since that time. Only the railroad siding remains to mark the site.

Columbia

DIRECTIONS: From Eureka, take U.S. 50 west for 15.5 miles. Exit left and follow for 14 miles. Exit left and follow poor road for 1 mile to Columbia.

Columbia was a classic "flash-in-the-pan" town, being discovered in spring 1869 and lasting just into early 1870. A small group of prospectors from Eureka discovered a ledge of very valuable silver ore. The resulting rush brought considerable activity to the area. A store opened and several cabins were built. The ledge, however, quickly disappeared and the small town vanished. Today only scant wood rubble shows that Columbia ever existed.

Corwin

DIRECTIONS: From Grubb's Well, continue southeast for 3 miles to Corwin.

Not much is known about Corwin. It was located along the old Overland Stage and Mail Company route, although it did not come into being until after the stage line had ceased operations in 1869. A post office was established on December 6, 1878, with John Chamberlain as postmaster. The office did not last even a year, closing on October 31, 1879. The only remains at Corwin are the collapsed ruins of two log cabins.

Deep Wells

DIRECTIONS: From Alpha, continue north on Nevada 278 for 7 miles to Deep Wells.

Deep Wells was a stop on the Eureka and Palisade and the Eureka-Nevada railroads. However, activity was present long before the railroad came through in 1874. In the late 1860s, a pioneer, taking advantage of the rush to Eureka, dug a well here to supply water for freight teams heading to Eureka. The enterprising pioneer charged $1 for watering an eighteen-mule freight team or two bits (25 cents) for four skins full of water. Once the railroad came through, a small camp formed around the still-active wells. Even with the railroad, some freight teams still worked between Palisade and Eureka, creating a market for the precious water. The water was also a necessity for the hot railroad engines as they made their way across the dry valley. During the 1870s and 1880s, Deep Wells had a steady population of about 25 and supported a store and a restaurant. Once Eureka faded in the 1890s, the residents left, and soon the site was used as just a railroad water stop. Today only a collapsing windmill marks the spot.

Diamond Springs (Diamond City)

DIRECTIONS: From Eureka, take Nevada 228 north for 29.5 miles to Diamond Springs.

Diamond Springs served as a Pony Express station during 1860 and 1861. Sir Richard Burton visited the station on October 9, 1860, and reported that it was run by an unfriendly Mormon couple. After the passing of the Pony Express, the station was used by the Overland Stage and Mail Company and also served as a telegraph station. George Francis Cox ran both of these enterprises. After the Overland Stage stopped running, the station house was left to decay.

Nearby, the small town of Diamond City formed during the mid-1860s. Initial silver discoveries were made in May 1864. However, production didn't begin in earnest until 1866. The primary producer was the Champion Mine,

right: The Diamond Springs charcoal kiln shortly before its collapse. (Northeastern Nevada Museum)

below: The Diamond Springs kiln as it is today.

All that remains of the Diamond Springs station.

and a small smelter was erected in 1873. Other mines included the Mammoth, Cumberland, Silver Wreath, Utah, and Cash. A post office, called Diamond, opened at the small town on September 3, 1874. While most mining in the area ended in the late 1870s, the charcoal kiln kept a few residents employed. The post office closed on July 10, 1884, and the town was abandoned soon after.

Today, mill foundations, stone ruins, and the charcoal kiln remain at Diamond City. At Diamond Springs, the newer Overland station remains. Beware, however; a swarm of mean bees that now inhabit the chimney can deter the visitor from getting a closer look! The setting is beautiful, surrounded by trees. Very enjoyable to visit, Diamond Springs and Diamond City are a must for both their history and their beauty.

Eureka (Napias)

DIRECTIONS: *Located on U.S. 50.*

The narrow canyons curving through the area that was to become Eureka were undisturbed by humans until September 1864, when a five-man party staked numerous promising claims in the district. The five men—W. O. Arnold, W. R. Tannehill, G. J. Tannehill, I. W. Stotts, and Moses Wilson—

Supplies being brought down Main Street, Eureka, in 1870. (Special Collections, Library, University of Nevada, Reno)

sold their claims to a New York mining company soon after. Activity was limited during the next two years, although small shipments of ore were sent to mills at Austin. Because of promising ore values, new prospectors came to the area in 1866. Ore from these claims, however, had a heavy lead content, which caused many smelting problems and held back any substantial production. It wasn't until early 1869, when Colonel G. C. Robbins completed the first successful draft furnace, that it became possible to mine large amounts of the silver-lead ore. The Robbins furnace treated ore from the Kentuck and Mountain Boy mines. In May 1869 another furnace, built by C. A. Stetefeldt and owned by Morris, Monroe, and Company, began operations. The furnace processed ore from the Champion, Buckeye, Grant, and Eureka mines. As the furnaces became functional, increasing the companies' ability to produce, people began to drift to the now-promising mining district. By October 1869, around 100 residents had settled in Eureka. The small camp acquired a post office, with George Haskell as postmaster, but had no name. Haskell christened the camp Napias, which is Shoshone for "silver." Within a short time, however, the town was renamed Eureka.

The main mines of the district were located on nearby Adams Hill and McCoy Hill. The town grew up in the canyon to the northeast. In the spring of 1870, Woodruff and Ennor set up a stage route. The line ran from Hamilton to

Eureka, Nevada, 1870. (Special Collections, Library, University of Nevada, Reno)

Eureka, and then to Palisade. Eureka began to grow quickly, and by October 1870, 2,000 people populated the town. By 1872, Eureka had grown to 4,000; by 1874 to 6,000; and by 1878, the town reached its peak of 9,000. In 1871 the first church was built in Eureka, and an official post office was organized. The Eureka Water Works built a $10,000, 55,000-gallon water tank, filled by water piped in from McCoy's Springs. The town at its peak was the scene of incredible activity: 125 saloons, 25 gambling houses, and many business establishments. Eureka also had 5 fire companies: Eureka Hook and Ladder, Rescue, Knickerbocker, Nob Hill, and Richmond Hose. In 1879, a $53,000 county courthouse was constructed. A landmark day in Eureka was Friday, October 22, 1875. After many delays the first train of the Eureka and Palisade Railroad, nicknamed the *Slim Princess,* arrived in town. Railroad access signaled the beginning of easy ore transportation for the mining companies and travel for the growing populace. The completion of the railroad made Eureka the center of stage transportation for central Nevada. Eureka became the distribution point for goods throughout the area.

Literary enlightenment came to Eureka on July 16, 1870, when the *Eureka Sentinel* began publication. Throughout Eureka's long history, the town was served by six newspapers, but the *Sentinel* remained the most important. The *Sentinel,* originally a weekly, was founded by Abraham Skillman and Dr. L. C.

Picking up supplies at the Eureka and Palisade Railroad depot. (Bancroft Library)

McKenny. On September 29, 1870, Fred Elliot and George W. Cassidy bought the paper. (Cassidy had run the *Inland Empire* at Hamilton before coming to Eureka.) The *Sentinel* was constantly being hit by natural disasters throughout its existence. On November 20, 1873, a fire caused $12,000 worth of damage. In July 1874 a flood wiped out the newspaper. In April 1879 another fire destroyed it. The *Sentinel* survived all of this adversity, however, and is the only paper still existing in Eureka. Another influential paper was the *Eureka Daily Republican,* founded by John C. Ragsdale. The *Daily Republican* began publication on January 4, 1877. On March 3, 1878, the paper was leased to Alfred Chartz, W. W. Wats, and Arthur McEwen. The paper had a dubious reputation because of personal attacks made by the publishers. One person in particular, Edward Ricker, took exception to a series of witticisms aimed at him, and on June 16, 1878, Ricker confronted Chartz. An argument broke out and Chartz ended up shooting Ricker. The uproar that followed the shooting eventually led to the closing of the paper on June 24, 1878. The next day, the *Eureka Daily Leader* began publication. The *Leader* was essentially the same paper. It was run by Canfield and Fish until October 1881, when James Anderson sold the *Ruby Hill Mining News* and bought the *Leader.* Soon after Anderson purchased the paper, he was shot and killed. People quietly talked about a curse at the paper. The *Leader* continued feebly until May 16, 1885, when it folded. The two remaining newspapers, the *Cupel* and the *Republican Press,* were short-lived. The *Cupel,* a Republican daily, was founded by William Taylor in March 1874. The name the *Cupel* came from the instrument used in assay work. The flood of July 24, 1874, destroyed the *Cupel,* and that newspaper was never heard from again. The *Republican Press,* of which little is known, started publication on November 30, 1884 and quickly faded into the literary sunset on May 9, 1885.

During the late nineteenth century Eureka fell victim to a series of disasters. There was a minor fire in 1872 that damaged only a few buildings. However, this served as a warning for the future. On March 23, 1875, fire once again visited Eureka. This time it started in the Montana House and quickly spread, destroying ten buildings and causing $25,000 in damage. Another, more serious, fire occurred on April 19, 1879, starting in Bigelow's Opera House and spreading throughout the town. The *Eureka Sentinel* building also burned, but the publishers, covered with wet blankets, put out a special edition about the fire before abandoning the building. The 1879 fire caused more than $1 million in damage. Only a little more than a year later, on August 17, fire once again struck, sweeping through the residential area of Eureka and destroying 300 homes.

Eureka was not exempt from natural disasters either. On July 24, 1874, a flash flood roared down the canyon, destroying 30 houses. The flood killed seventeen people and caused $100,000 in damage.

One of the many burro teams that brought charcoal to the Eureka smelters, 1902. (Northeastern Nevada Museum)

By 1870 the mines throughout the Eureka District were starting to achieve prominence, leading to a rapid increase in the number of smelters in the town. Once a successful type of smelter was devised, the smoke-belching monsters sprang up everywhere. In 1870 David Buel and John Bateman, after buying the Champion and Buckeye mines, built two furnaces at the lower end of town (which were later sold to the Eureka Consolidated Mining Company). Also in 1870, the Jackson Consolidated Mining Company built two furnaces at the old Wilson Furnace site. The J. J. Dunre Company bought the H. P. Nevin Furnace site at the south end of town and constructed a furnace to treat ore from the Richmond Mine. By 1874, 24 furnaces were in operation in Eureka. All the heavy smelter smoke cast a constant pall over the town, and Eureka soon earned the moniker of "the Pittsburgh of the West." The four largest companies in the district at the time were Richmond Consolidated (6 furnaces, 360 tons capacity), Eureka Consolidated (5 furnaces, 300 tons), Atlas (2 furnaces, 120 tons), and Ruby Consolidated (2 furnaces, 100 tons). While smelting was the primary method of extracting the silver and lead, there were also a few stamp mills. These mills processed ore from mines in the Eureka District, the nearby Diamond District, and the Newark Valley (White Pine County). The largest of the mills was the Lemon Mill, moved from Shermantown (White Pine County), where it was known as the Metropolitan Mill, and put into operation in Eureka in 1872. The Lemon had fifteen 1,000-pound stamps, six amalgamating pans, and two furnaces. However, because there wasn't a consistent flow of ore, the mill was operated only intermittently.

The day shift at the Windfall Mine, 1912. (Northeastern Nevada Museum)

During the early 1870s, Eureka was the scene of intense mining activity, with many mining companies active in the district. While Eureka Consolidated and Richmond Consolidated were easily the most important, several smaller companies also had a significant impact on Eureka's production figures. Among them were the Buttercup Company (which ran the Miner's Dream, Prosper, Kentuck, Mountain Boy, Fairview, and Madrid mines), Silver Lick Consolidated (Vera Cruz, Silver Lick, California, Body Burns, and Black Chariot mines), Lander Consolidated (Banner, San Jose, Charleston, Danube, and Republic mines), Ruby Consolidated (Bullwacker, Dunderberg, Lord Byron, Valentine, and El Dorado mines), Mountain View Consolidated (Anna, Mountain View, and Mazeppa mines), Alkali Mines Company (the Windfall Mine), and Phoenix Silver Mining Company (Adams, Empire, and Lexington mines). However, by the late 1870s, virtually all of the smaller companies had been absorbed by either Eureka Consolidated or Richmond Consolidated. Then, during 1877, these two companies became embroiled in a hotly contested legal suit. Both companies claimed a rich body of ore known as the Potts Chamber. Richmond Consolidated had followed an ore shoot that led to the Potts Chamber. However, in the process, the company had crossed a compromise plane that had been set up by the two companies. Richmond thought it had the right to follow the ore shoot. The courts thought otherwise and Eureka Consolidated won the case, but not until 1881. Richmond Consolidated had to reimburse Eureka Consolidated for the ore that had been mined from the Potts Chamber, at an estimated value of $2 million.

Although water began to flood some of the mines in 1881, Eureka continued to lead Nevada in silver-lead ore production. But the end was approaching. By the middle 1880s, the ore bodies were almost exhausted. To compound the problem, the price of silver dropped dramatically. Both Eureka Consolidated and Richmond Consolidated struggled on into the 1890s but showed little profit. Richmond folded in late 1890 and Eureka in 1891. The Eureka District was extremely quiet for about a decade. Mining activity was almost nonexistent, and the population of Eureka plummeted. Many businesses left, and on June 13, 1900, the Eureka and Palisade Railroad ceased operation.

In 1905 a small revival took place. In September the Richmond-Eureka Mining Company was formed, with W. G. Sharp as president. The company showed fair profits, with its best year being 1909, when the mines produced $711,000. But the railroad to the mines was washed out in 1910, and the company soon gave up the effort. After that, major mining activity ceased and there were only occasional efforts to reopen the mines. Small revivals occurred in the mid-1920s and the late 1930s. All told, the Eureka District produced ore valued at a little over $108 million, with the bulk ($60 million) being lead.

The Richmond-Eureka Mining Company Mill in 1908. (Northeastern Nevada Museum)

Although mining had almost completely stopped, the Eureka and Palisade Railroad was revived in May 1912. The railroad was renamed the Eureka-Nevada Railroad. Though the Eureka District itself was quiet, the town was still a supply center for newer mining districts in the area. The railroad continued to run until September 1938, when it was finally abandoned. Since that time, Eureka has been a peaceful town with a fairly steady population of about 200. The town has recently experienced a revival on two fronts: A few of the mines have reopened and have helped bring the town back to life. Also, many of the beautiful and historic buildings are undergoing renovation, and the results are fantastic. Among the best buildings to see are the courthouse, the *Eureka Sentinel* building (which houses a small museum), the old movie theater, the post office, and the churches. Also a must are the nine cemeteries located around Eureka. All possible supplies and accommodations are available in the town. With its rich history, Eureka is one of the more interesting places in the state—definitely a must for any tour of Nevada.

Evans

DIRECTIONS: *From Hay Ranch, continue north on Nevada 278 for 7 miles to Evans.*

Evans was a siding on the Eureka and Palisade and the Eureka-Nevada railroads, located near the Evans Ranch, owned by J. V. Evans. Nearby ranchers used the siding to ship their goods to Palisade and Eureka and also to receive supplies. While the population of Evans varied, it never had more than 25 residents. The railroad folded in 1938, but Evans continues to operate as a ranch. Most of the buildings at the ranch are from more recent times, although some of the original structures still exist.

Gerald

DIRECTIONS: No roads go near this site now. Located 3 miles southwest of Palisade, alongside railroad tracks.

Gerald was a stop on the Central Pacific and the Western Pacific railroads. A small camp formed here during the mid-1880s. Many miners who worked in the nearby Safford District lived here. A post office opened on July 24, 1882, with Edwin Clay as postmaster. The office remained in operation until April 30, 1887. Only some wooden rubble now marks the site, and the long walk to Gerald is not worth the effort.

Goldville (Leeville) (Lynn District)

DIRECTIONS: From Carlin (Elko County), follow Maggie Creek Road north for 17 miles to Goldville.

The Goldville District was discovered in April 1907, by Joseph Lynn and William Barney. The two men's big find was the Lynn Big Six Mine. During the mine's first year of operation, it shipped $21,000 worth of ore to Salt Lake City. This success spawned a small rush to the new district. The Lynn Big Six was the only mine of any size in the district, however; all of the other ore found in the area occurred in placer deposits. By the end of 1908 these placer deposits had been exhausted, and the 20 or so prospectors living in Goldville had left.

But that was not the end of Goldville. In 1912 the Lynn Big Six Mining Company, with Henry Hanker as president, reopened the Lynn Big Six Mine, and the company was soon shipping ore to Salt Lake City at a price of $150 a ton. The mine itself was 200 feet deep, with substantial drifts at the 100-foot and 200-foot levels. A small camp soon reformed, and a post office opened on July 22, 1913, with William Barney, one of the original founders, as postmaster. Though the post office closed on August 31, 1917, the Lynn Big Six Mining Company was still quite active. In 1917 the company built an amalgamating mill, originally with a 10-ton capacity but later increased to 25 tons. The company continued to expand, and by 1922 it had twelve claims and controlled 220 acres. Besides the Lynn Big Six Mine, the company had developed a new 800-foot tunnel mine named the Gold Dollar Mine. The capacity of the mill was enlarged to 50 tons in 1919 and to 100 tons in 1921. The value of ore from the mines was very inconsistent, however, ranging from $2 to $300 per ton. Total production from the Goldville District from 1909 to 1921 was $100,000.

The company continued to expand, although the work force remained at about 15. By 1925 the company had staked 36 claims covering 420 acres. While the Lynn Big Six had been expanded only to the 300-foot level, the Gold Dollar was extended to more than 2,000 feet. However, the company gave up the district in May 1926, after producing $225,000. The district was quiet until 1937, when the property was taken over by the Beaver Crown Consolidated Mining Company, with R. M. Holt as president. This company dug three new tunnel mines: Talbot (1,400 feet), Mill (800 feet), and Bull Moose (700 feet). Ore values, however, were quite poor, and by 1939 the company folded. Today remains are scant, with the only ruins of recognizable origin being the mill foundations. Mining activity in the area has increased recently, though, especially in the nearby Maggie Creek District. The Newmont Gold Company has had control of the district's claims since the 1960s, and the Lynn District is the core of a series of immense open-pit mines. Microscopic gold is now the main product, and Newmont's operations have been consistently producing more than one million ounces of gold per year. With ore deposits believed sufficient for twenty years, the Lynn District will be active for a long time to come. The gold mines in the district are among the top five gold producers in the country and are an economic mainstay of Eureka and Elko counties. Tours of the mining and milling complexes, available by contacting Newmont in Elko, are well worth the time.

Goodwin

DIRECTIONS: From Roberts Creek, follow old Pony Express trail southwest for 6 miles to Goodwin.

Goodwin was located at a small group of springs in the dry flats west of Roberts Creek. This site served as a stop for Pony Express riders and the Overland Stage. Though the spot was not an official Pony Express station, riders often stopped to water their horses here before continuing to the next water stop, at Grubb's Well. One small cabin was built at Goodwin, and only a few old logs from that structure now mark the site.

Gravelly Ford

DIRECTIONS: Located 2 miles east of Beowawe.

Gravelly Ford is more of a historical site than a town, although a post office did function here from February 22, 1869, until January 31, 1870.

It is alleged that here, long before any settlement formed, James Reed, a member of the ill-fated Donner Party, killed John Snyder, also a member. However, there is controversy as to whether Gravelly Ford was really the site of the murder. Some accounts place the event at Iron Point, near Battle Mountain.

Probably the most interesting and controversial part of Gravelly Ford's history is the Maiden's Grave. There are two stories about the origin of the Maiden's Grave. The more popular version is that a young girl, Lucinda Duncan, got sick, died, and was buried in the valley along the Emigrant Trail, near Gravelly Ford, heading toward Beowawe. When the Central Pacific Railroad was being built, it was discovered that the tracks would destroy the grave. The body was moved and re-interred a short distance away on a small hill. The new grave was marked by a large wooden cross. Members of the railroad building crew began to maintain the grave, and a tradition was born. The controversy centers on the age of the deceased woman. The above version says she was fourteen, while the version related by Iva Robertson Rader, granddaughter of Lucinda Duncan, says she was seventy! Mrs. Rader states that her grandmother was born in 1793 in Virginia and died at Gravelly Ford of an aneurism of the heart, with most of her children and grandchildren standing nearby. I leave it up to the reader to decide which version is more plausible.

Gravelly Ford also had a history of conflict with Indians. Twenty-three immigrants were murdered nearby during the Indian troubles of the 1850s and 1860s. A small camp did form at Gravelly Ford in the 1860s during the construction of the Central Pacific. In 1880 the camp had a population of 42 and supported a store, a restaurant, and a telegraph station. Today only wood scraps and depressions mark the site.

Grubb's Well (Camp Station)

DIRECTIONS: *From Bar, head south for 11 miles. Exit right onto old Pony Express road (very poor) and follow for 6.5 miles to Grubb's Well.*

Grubb's Well was a short-lived Pony Express station, established in August 1861. A small ranch was active here at that time and was pressed into service during the summer. However, the time and scope of its activity was quite limited. Later, after the failure of the Pony Express, Camp Station was renamed Grubb's Well and was used by the Overland Stage until 1869. Today a lot of dilapidated buildings remain at Grubb's Well, but all were built around the turn of the century. None could have served as the station. The buildings were used as a ranch until the Great Depression of the 1930s, when the property was abandoned for good.

Hay Ranch

DIRECTIONS: From Palisade, head south, then east on poor dirt road. Exit south onto Nevada 278 and follow for 15 miles to Hay Ranch.

Hay Ranch has served many purposes during its long history. First it operated as a stage station during the 1860s, while growing 1,000 tons of hay each year. The hay fed teams from as far away as Pioche. When Eureka began to boom, Hay Ranch became an important food stop for stages and freight teams heading south. A boardinghouse, a restaurant, and a livery stable thrived during the late 1860s and 1870s. On March 29, 1872, a Palisade-Eureka stage was robbed near here, providing Hay Ranch with a bit of history that is rare in Eureka County.

Hay Ranch achieved even more importance once the Eureka and Palisade Railroad was completed. During the construction of the line, railroad executives purchased 2,500 acres of nearby bottomland. This area was fenced in, and more than 300 mules were used to harvest the hay in the fall. Once the railroad was completed to Eureka in 1875, Hay Ranch's population stabilized at about 25 people. This figure has remained fairly constant through today, and the ranch is still active. Only a couple of vintage buildings remain. One of them, obviously moved from the old railroad bed, appears to have been the Hay Ranch station house.

Hot Springs (White Sulphur Springs)

DIRECTIONS: From Bruffey Ranch, backtrack and head north for 2 miles to Hot Springs.

Hot Springs was a short-lived luxury spot that existed while nearby Mineral Hill was successful. The springs were bought by a Dr. Davenport in March 1871. The water was very high in mineral content and reputedly did wonders to help diminish the suffering caused by assorted diseases. Davenport built a large bathhouse at the springs and renamed them White Sulphur Springs. He advertised heavily in the *Eureka Sentinel* and apparently did a fairly brisk business. Once Mineral Hill began to fade, however, business slowed. As near as one can tell, Davenport closed up shop in the mid-1870s, and no other interest emerged in the area. Today absolutely nothing remains of the wooden bathhouse, but the hot springs are still going strong.

Keystone

DIRECTIONS: From Bar, continue north for 4 miles to Keystone.

Keystone was an obscure mining camp that was the scene of some short-lived and low-key activity. The Keystone Mine was originally discovered in the late 1860s by one of the many wandering prospectors who were drawn by the Eureka boom. But it wasn't until 1898 that any measurable activity took place. At that time R. D. Clark dug shafts of 150 feet and 105 feet, uncovering some ore that assayed as high as 600 ounces of silver and $120 in gold per ton. Even though the sample had been a carefully selected shipment of ore, caution was thrown to the winds. A small concentrator was built at the mine. Other plans included the immediate construction of a smelter. A camp of 50 sprang up, and on April 14, 1898, a post office opened. Unfortunately, that first load of ore contained everything of value in the mine. Efforts to locate new deposits were fruitless, and by the end of the summer most of the population had already left. The post office closed on September 27. Soon the only signs of the camp were the concentrator and the partially completed smelter—fitting memorials to the follies of man. Only rubble marks the site today.

Mill Canyon (Majestic Camp)

DIRECTIONS: From Cortez (Lander County), head north on Nevada 306 for 4 miles. Then exit right and follow rough road for 3 miles. Exit right and follow this road for 4.5 miles to Mill Canyon.

The Mill Canyon Mining District was organized in 1863 as a result of mining activity at nearby Cortez. Simeon Wenban, the owner of most of the mines in Cortez, built an 8-stamp mill here in June 1864. Delivery of the ore to the mill was extremely difficult. Ore was brought over the mountains by mule or by large ore wagons, which had to take an even longer route. In 1869 the mill was enlarged to 16 stamps, and four roasting furnaces were added. All of the claims in Mill Canyon were purchased by Simeon Wenban in 1867, but it wasn't until the 1870s that the mines were extensively explored. Among the most important mines were the Cynthia (the first mine in the district), the Empire State (discovered in 1872, produced $2,500), the Hidden Treasure ($35-per-ton ore), the Falconer ($35-per-ton ore), and the Berlin (produced $40,000). Overall, however, these mines were not successful, and after 1873, most of them were abandoned. The mill ran until 1886, when it closed after construction of a new mill at Cortez.

The district remained quiet until 1909, when a few of the mines were reopened after Patsy Clark, a prominent Western mining man, began to show interest in the area. A gravity concentrator was built, and it treated ore from the Edwin and Falconer mines. But the effort was unsuccessful and was given up in 1910. It was 1928 before another company ventured into the district. The Majestic Mines Corporation was incorporated in 1928. Before the company sold its holdings to Belle McCord Roberts and M. J. Hough in November 1929, $200,000 was produced. These two men organized the Roberts Mining and Milling Company, which built a 25-ton cyanide mill that operated until early 1938. More than 30 men were employed at the mill, but most lived in nearby Cortez. When the company folded in 1938, the district was abandoned for good. The remains in Mill Canyon are spread out. The most visible ruins are those of the Roberts Company mill at the mouth of the canyon. Stone ruins are scattered along the canyon. The road to the site is quite dangerous; travelers should exercise caution.

Mineral

DIRECTIONS: *Located 4 miles west of Mineral Hill, on Nevada 278.*

Mineral was an important stop on the Eureka and Palisade and the Eureka-Nevada railroads. It served as the dropping-off point for all freight and passengers bound for Mineral Hill. During the 1870s, a substantial settlement developed here as Mineral Hill continued to grow. The stop had the distinction of being the only regular eating station between Eureka and Palisade. The owners of the restaurant charged $1 per meal and reportedly served heaping portions. Besides the restaurant, a small saloon, many tents, and a small railroad depot also occupied the area along the tracks. When Mineral Hill began to decline in the mid-1870s, only a handful of people stayed there. Soon there were none. Today only mounds reveal where Mineral Station once stood.

Mineral Hill

DIRECTIONS: *From Alpha, head north on Nevada 278 for 11.5 miles. Exit right and follow poor, sandy road for 4 miles to Mineral Hill.*

Two disgruntled prospectors, John Spencer and Amos Plummer, after leaving the Reese River District, discovered some rich silver float here in June 1869. They staked claims and formed the Mineral Hill Mining District. A test load of ore sent to Austin yielded $200 per ton and showed values in gold, silver, copper, zinc, and lead. Several mines were started, with the most valuable being the Austin, Mary Ann, Rim Rock, Grant, Star of the West,

The 15-stamp Mineral Hill Mill, 1871. (U.S. National Archives)

Vallejo, and Pogonip. The pair sold their holdings in 1870 to George Roberts and William Lent of San Francisco for $400,000. Roberts and Lent immediately organized the Mineral Hill Mining Company and constructed a 10-stamp mill, built by Huber and Curtis. The mill contained six Wheeler amalgamating pans, three settlers, and a Stetefeldt furnace. The mill was started in September 1870 and soon enlarged to 15 stamps. Besides the Mineral Hill Mining Company, many other companies formed and were working in the district by 1870. The Austin Mining Company owned the Austin and Western Slope mines. Its ore was sent to the Auburn Mill (Reno) and the Manhattan Mill (Austin). The Robinson, May, and Company (Big Sandy Mine) and the Grass Valley Tunnel Company were also active.

Interest in the Mineral Hill District grew rapidly, and the area experienced a small rush during the summer and fall of 1870. By fall, more than 400 people called Mineral Hill home. The town boasted four saloons, two hotels, a Wells-Fargo station, and ten other business establishments. The year 1871 brought about many changes. Early in the year, a Colonel Hyman completed the three-story Grand Hotel, and soon after that, the Mineral Hill schoolhouse was also completed. On May 9, a post office opened up, with Thomas J. Isabell as postmaster. Most of the mining action in Mineral Hill during 1871 concerned English interests. In May, an English corporation bought the holdings of the

Mineral Hill Mining Company for $1.2 million. A month later, the corporation, headed by Albert Grant, resold the holdings to another English company, also headed by Grant, for $2.5 million! The company organized as the Mineral Hill Silver Mining Company, Limited. The company had a working capital of $1.5 million and owned 41 mines. The richest of these mines were the Giant, the Star of the West, and the Troy. Besides running the 15-stamp mill (the Atwood), the company quickly built another mill, the 20-stamp Taylor, at a cost of $135,000. The latter mill was built by H. W. Bordwell, who also built the International Mill in Eberhardt (White Pine County). The Taylor was first fired on December 28, 1871. Unfortunately the new mill was not very successful and remained in operation for only about four months before it was sold to the Leopard Mining Company of Cornucopia for $17,500. Both mills had originally been fitted with Stetefeldt furnaces, but given the value of the ore, they were too expensive to operate, and the mills were changed to dry crushing and raw amalgamation facilities in 1872.

During 1871 and 1872, the company mined the area at an accelerated rate, and more than $1 million were produced. This included one single ore shipment in 1871, via Wells-Fargo, of $701,000. By summer 1872 the company had mined out all the ore. It folded before the end of the year, thus terminating the employment of almost 100 men and leading to an exodus from Mineral Hill. A large fire also helped to chase people out of the town. On June 17, 1872,

The Mineral Hill Mining Company's main office, 1908. The men in the photo are (left to right): John Hadon, Frank Winzel, George Isaacs, Frank McDonough, and Bert Bruffey. (Nevada Historical Society)

The main ranch house at Bruffey Station, located on the stage route between Mineral Hill and Union.

a fire started in King's Hotel and Thomas J. Isabell's store. While no lives were lost, only five businesses survived. Little more than a week later, on June 26, the Woodruff and Ennor stage was robbed of $5,000, the entire payroll for the mine workers. By spring 1873, only a handful of people remained. The company tried for two years to discover new deposits, but all efforts were in vain. The Mineral Hill Silver Mining Company went bankrupt in 1874, leaving many disgruntled investors who never received a dividend. Mining in the district was virtually nonexistent until 1880. All Mineral Hill Company holdings were sold at a sheriff's sale, and the Austin and Spencer Company bought the properties and began to rehabilitate the mines. At the time only 18 miners remained in Mineral Hill. A year later, however, 150 people were back in town. Production was quite slow: Only 113 tons of ore were produced from 1880 to 1882. But in 1883 the company had a breakthrough and discovered a new ore deposit. The old Atwood Mill was restarted, and the Austin and Spencer Company experienced a small boom until 1887. Then the ore ran out and the company quietly folded. While the mines were leased occasionally, the district was essentially dead for the next fifteen years. The post office closed on July 19, 1890, and didn't reopen until February 6, 1902.

In 1902 a couple of small companies began leasing the claims on Mineral Hill. While activity in the district increased, real production didn't begin until 1910. In 1906 the Mineral Hill Consolidated Mines Company was incorporated in Arizona. The company, headed by S. C. Pratt, gained control of eleven claims, reopened several old mines, and did 2,000 feet of underground development before beginning production. A small cyanide mill was built in 1911 to treat new ore and to reprocess ore on the old dumps. The company experienced a small surge of profits until 1915, when ore values declined. The post office closed for good on April 15, 1914, and Mineral Hill began to drift toward ghostliness. Mineral Hill Consolidated continued to do some very limited work in the district. By time it folded in 1920, only 10 people were left in Mineral Hill. Total production from 1905 to 1920 was only $360,000. The production from the district as a whole was just under $7 million. The district remained dormant, except for minor activity in the 1930s, until the 1980s, when a new company began to conduct cyanide leaching operations on the extensive ore dumps near the town. Very little remains of the town itself. Some stone ruins are still there, along with rubble from several wood structures. The most impressive ruins are the extensive mill remains. The road to Mineral Hill is very sandy and treacherous. Use extreme caution. Remember, sometimes a hike is better than a drive!

Mount Hope (Morlath) (McGeary) (Garden Pass District)

DIRECTIONS: From Garden Pass, continue north on Nevada 278 for 2.5 miles. Exit left and follow for 3 miles to Mount Hope.

The initial discoveries at Mount Hope were made in the fall of 1871 by a man named McGeary, who organized the McGeary Mining District. By January 1872, twenty claims were being worked. A small camp soon formed and a townsite was laid out by Mike Browsky and Company in early 1872. By summer, several mines had been developed, including the Mexican, Josephine, Black Cloud, Little Frank, Geneva, Rambling Boy, Phillip Sheridan, Champion, and Badger State. The richest of them was the Mexican, with ore values running at around $100 per ton. The Mount Hope mines attracted a lot of attention from Eureka's mining companies. George Bibbins, superintendent of the Magnolia Mining Company, did a complete prospecting canvass of the district and found the mines to be very promising. Bibbins, along with Marceliena Carabantes, discovered the Mexican and Carabantes Mine, which brought a fair profit to the pair.

In August 1872 the district was renamed the Hope Mountain Mining District because the mines were located on the slopes of Mount Hope. During that month a rich new mine, the Last Rose of Summer, was discovered by Captain John M. Foley. The mainstay of the camp, however, became the Josephine. By April 1873, the mine was more than 500 feet long and was producing $125-per-ton ore. There were rumors during the summer of 1873 that President Ulysses S. Grant was going to visit the camp on his Western trip, but the visit never materialized. After 1878 Mount Hope quickly died. The camp was then entirely deserted until 1886, when the Mount Hope Mine was reopened and operated off and on until 1947. The Callahan Lead-Zinc Company controlled the mine during the 1940s and produced several million dollars in zinc. In March 1947, fire forced the company to fold, destroying the mine workings, powerhouse, shop, compressors, and light and power units. The tragedy put 80 men out of work, and the district was abandoned. Interest in the district was revived when Exxon announced on August 7, 1978, that a large molybdenum deposit had been discovered. Exploratory drilling began in late 1978, and estimates indicate that the deposit could last for 50 years. Mount Hope could be around for a long time! However, a large drop in molybdenum prices forced Exxon to abandon its mineral operations in the 1980s. At the site, remains from the original Mount Hope settlement are scarce. Most of the visible remains are from leasing activities in the 1920s and 1930s and the extensive operations of the Callahan Company during the 1940s.

Nevin

DIRECTIONS: From Bar, continue north for 4.5 miles. Exit left and follow for 8 miles to Nevin.

Nevin was a very small, short-lived mining camp. Discoveries were made in the summer of 1906, and the first shipment of ore to Eureka showed fairly decent values. A small camp of 15 formed. On October 31, a post office, with Frederick Rocker as postmaster, opened. However, the ore deposit ran out after only a few months. On May 18, 1907, the post office was rescinded—before it ever became official. Only some scattered wooden rubble now marks the site.

Oak

DIRECTIONS: From Garden Pass, continue north on Nevada 278 for 11 miles to Oak.

Oak was a siding on the Eureka and Palisade and the Eureka-Nevada railroads, which served as a supply point for the ranches in Diamond Valley from 1875 until the railroad tracks were torn up in 1938. No settlement ever developed at Oak, and the site was abandoned when the railroad folded. Nothing remains at the site today.

Palisade (Palisades)

DIRECTIONS: From Beowawe, head east on poor road for 15 miles. Exit right and follow for 3 miles to Palisade.

Long before Palisade became an important railroad center, the area was the scene of immigrant activity. Palisade was located near a couple of major trails. The first visitor was Zena Leonard, who traveled through picturesque Palisade Canyon (named after a similar formation along the Hudson River in New York State) in 1833.

The town of Palisade came into being in 1868 and served as a stop on the new transcontinental railroad, the Central Pacific. The station quickly became prominent as the shipping point for supplies to mining districts in the eastern portion of Nevada. A post office, with John Merchant as postmaster, opened on May 2, 1870. In August 1871, the growing town suffered a setback when a fire destroyed the Sutherland block of the town and caused $50,000 in damage. Palisade recovered quickly and in 1874 grew in importance when construction of the Eureka and Palisade Railroad was begun. The town became the headquarters for the railroad and its 4 locomotives, 58 freight cars and 3 gaudy yellow passenger coaches. By 1878, more than 31 million pounds of base bullion had been shipped by the Eureka and Palisade Railroad, keeping the Palisade shipping companies extremely busy.

Eureka and Palisade Railroad Engine #5 at the Palisade depot. (Nevada State Museum)

Supplies in Palisade, waiting for shipment to Eureka. (Northeastern Nevada Museum)

A couple of fraternal organizations made their homes here. The International Order of Odd Fellows and Masons constructed beautiful lodges in the town. Catholic and Episcopal churches and a schoolhouse were built. The Eureka and Palisade Railroad Company built a large shop where the freight cars for the line were manufactured, thus saving a lot of expense for the company and employing some of Palisade's residents. While Palisade had a pouplation of close to 600 in the mid-1870s, by 1880 the town had settled down to a consistent population of 250. In 1882 a brand-new spacious railroad station was completed and served as the station house for both the Central Pacific and the Eureka and Palisade railroads. When Eureka declined, however, Palisade declined. Palisade was still the main shipping point for the mining districts to the south, but Eureka was the principal supplier of income for Palisade's businesses. As Eureka's mines slowed down, the Eureka and Palisade Railroad runs became more and more infrequent. In May 1912 the railroad was reorganized and renamed the Eureka-Nevada. In 1908, a third railroad, the Western Pacific, began operation, but its arrival did not really

The severe 1910 Palisade flood. (Northeastern Nevada Museum)

help Palisade. A series of disastrous floods struck the town in 1910, wiping out many of the businesses and damaging all three railroads. In 1915 Palisade still had a population of 242, but within a few years the figure had shrunk to less than 150. When the Eureka-Nevada pulled up its rails in 1938, the end of Palisade was in sight. During the following years, only a few families continued to live in the town, and the post office operated until May 26, 1961. Soon after that, Palisade became a ghost town for good. Today only two buildings remain in the town, both of them small wooden cabins. Parts of several stone buildings and wooden rubble are also left. An extensive cemetery remains on the west end of the townsite. The ride to the site is well worth the trip for the beautiful scenic views in Palisade Canyon.

Pine (Pine Station)

DIRECTIONS: *From Oak, continue north on Nevada 278 for 6.5 miles to Pine.*

Pine Station, outside of Alpha, was one of the most populous stations on the Eureka and Palisade Railroad line. The population often rose to more than 100, mainly because of the work force needed to produce charcoal for the Eureka smelters. A post office was active here from May 7, 1886, until February 14, 1888. Although Pine Station had a substantial population, nothing of consequence was ever constructed. A majority of the workers were transient, and most lived in tents. Only a few built cabins. Only one business, a combination saloon and store, ever opened. By the turn of the century, Pine had just about outlived its usefulness and was relegated to a railroad siding. Today nothing noticeable remains.

Pinto

DIRECTIONS: From Eureka, head east on U.S. 50 for 5.5 miles to Pinto.

Pinto was a milling camp rather than a mining camp. A 20-stamp mill, equipped with a Stetefeldt furnace, was constructed in 1871. Built by A. M. Scheidell and started on December 13, the mill processed ore from the Secret Canyon District and the Silverado (Pinto) District in White Pine County. A small camp of 15 people lived near the mill, and a school was built there. A man named Ben Levy opened a store in October 1871, and soon a few other businesses joined him. A post office, with Jimius M. North as postmaster, opened on August 6, 1875. The mill was leased by the Geddes and Bertrand Mining and Milling Company, active in the Secret Canyon area, in November 1876. The mill closed for good in 1884, and the post office followed suit on November 14 of the same year. The mill was dismantled and Pinto ceased to exist. Today only faint store ruins and hard-to-find mill foundations mark the site alongside U.S. 50.

Prospect

DIRECTIONS: From Eureka, take U.S. 50 east for 1.5 miles. Exit right onto New York Canyon Road. Follow for 2.5 miles to Prospect.

Prospect was a small settlement that formed in 1885, serving more or less as a residential camp for miners working in the nearby mining districts. While there were no prominent mines around the camp, a small pro-

Prospect, 1907. (Denver Public Library)

ducer was located on Prospect Mountain, overlooking the camp. A post office opened on March 3, 1893, with Earl Tremont as postmaster, to serve the 50 residents of Prospect. In 1908 a small smelter was built in Prospect, along with a couple of boardinghouses. The town boasted a saloon and a school. Supplies were brought here via thrice-weekly stage from Eureka. Once the mines near Prospect and along nearby Secret Canyon closed down, the camp quickly emptied. The post office closed on April 30, 1918, and soon the camp was abandoned for good. Several ruins remain in Prospect. Stone foundations and scattered wooden rubble mark the site. The ruins of the smelter, at the base of Prospect Mountain, are the most recognizable.

Raines

DIRECTIONS: *From Hay Ranch, continue north on Nevada 278 for 13.5 miles to Raines.*

Raines was a small siding on the Eureka and Palisade and the Eureka-Nevada railroads. Built in 1874, the siding was used until the rails were pulled up in 1938. No settlement ever developed here, since no activity other than ranching ever took place. The siding's name came from E. P. Raines, who ran the Raines Ranch, where the siding was located. The ranch is no longer active, but many interesting buildings remain. Ask for the owner's permission before exploring.

Roberts Creek Station (Willow Creek) (Sheawit Creek) (Leopold District)

DIRECTIONS: *From Eureka, head west on U.S. 50 for 15 miles. Exit right and follow for 13.5 miles to Roberts Creek Station.*

Roberts Creek was named for Bolivar Roberts, the division superintendent of the Pony Express. The station tender was Peter Neece. He had many conflicts with the Indians and killed two of them in one battle. Newspapers and magazines arrived at the station only twice a year, making the life of the tender a lonely one. John Fisher, later to become a prominent judge, was a Pony Express rider between Salt Lake and Roberts Creek. When the Overland Stage and the Pony Express stopped running, Roberts Creek became a successful ranching operation, and it is still active.

Some mining also took place near Roberts Creek. The Leopold Mining District was established by two men named Roberts and Tucker. In 1870 the men located the O'Dair Mine, which produced small amounts of ore with values in silver, lead, and copper until 1872, when this activity ceased. The last mining activity in the district took place in 1877. In April of that year a man named Gallagher worked two mines, the Levan and the Richmond #2, but very little production was realized. Soon the district was abandoned for good. Nothing marks the site today. At Roberts Creek, nothing from the Pony Express era remains, although some old log structures are left. These remains are from a later period, possibly the Overland Stage era of the 1860s.

Romano

DIRECTIONS: *From Garden Pass, exit right on sandy road heading into Diamond Valley. Follow for 7 miles to the Romano Ranch.*

Romano never was more than a ranching settlement, and its peak population was fewer than 20. The ranch, however, was very important to Diamond Valley. A post office opened here on February 24, 1902, with Frank Romano as postmaster, and remained functional until August 31, 1914. The post office reopened on February 3, 1919, but closed for good on February 15, 1929. The post office served as the mail depot for all the ranchers in Diamond Valley, thus saving the residents the trip into Eureka to get their mail. The ranch continued to operate after the post office closed and was abandoned only within the last twenty years. The main ranch house remains, and although run down, it still presents some beautiful lines. The view of Diamond Valley is impressive, and a visit to the site will be quite enjoyable.

Ruby Hill

DIRECTIONS: *Located 2.3 miles southwest of Eureka.*

Ruby Hill's rich mines were first discovered in 1869 by Owen Farrel, M. G. Clough, and Alonzo Monroe. The true discoverer, however, was an Indian who led them to the strike site. The men located two rich mines, which they named the Buckeye and the Champion. By the early 1870s, a camp began to form around the mines. As more mines were discovered, the camp continued to grow. A post office, with Josiah W. Jones as postmaster, opened on September 23, 1873. By that time several large mining corporations were working the Ruby Hill District. Eureka Consolidated owned the Buckeye, Mammoth, and Sentinel mines. Richmond Consolidated ran the Richmond, Tip Top, Lookout, and Victoria mines. K & K Consolidated and Jackson Consolidated also were active. In 1875 the Ruby Hill Railroad, known as the "3 × 3" because it was three miles long and three feet wide, was completed. The railroad's availability allowed easy transportation of ore to the

Downtown Ruby Hill. (Special Collections, Library, University of Nevada, Reno)

smelters in nearby Eureka and also aided the rapid expansion of mining in Ruby Hill. In fact, the Ruby Hill Railroad had the first locomotive in Eureka. The Eureka and Palisade construction was still stalled in Alpha. While designed mainly as a conveyance for ore, the railroad was such a convenience to the people that a small passenger coach was built and attached to the end of the normal load of six 10-ton ore cars. The railroad continued to be called the Ruby Hill, even though the Eureka and Palisade bought the line when its own line was completed to Eureka.

Ruby Hill was the scene of many heated mining disputes. The first, over the Lookout Mine, took place in 1872 and involved Eureka Consolidated and Richmond Consolidated. At one point, Richmond Consolidated placed armed guards around the mine. A compromise was reached in June 1873, before actual violence broke out. Richmond agreed to pay $85,000 to Eureka Consolidated in exchange for full possession of the mine. Part of the compromise was that Richmond Consolidated would agree to stay within the boundaries of the claim area, both above and below ground. In 1877 the company broke that agreement by following an ore shoot from the Lookout Mine to the Potts Chamber, a rich silver deposit. Four years of litigation followed, and finally Eureka Consolidated prevailed. Richmond was forced to pay $2 million to Eureka for ore removed. In the late 1880s, Richmond Consolidated became embroiled in more litigation. The company accused the owners of the Albion Mine of trespassing. The owners of the mine lost the case, and in 1894 they were forced to close the mine and its 100-ton smelter.

By 1878 Ruby Hill had reached its peak population of 2,500. Two newspapers served the town during its boom years. The *Ruby Hill Mining Report*, run by Mark Musgrove, former owner of the *Belleville Times*, began publication on October 17, 1878. This paper, however, was overshadowed by the popular *Eureka Sentinel* and folded in the spring of 1879. Another newspaper, the *Ruby Hill Mining News*, began publication on April 24, 1880. The owner, John Anderson, ran the publication until December 1884.

By the middle of the 1880s, mine production began to fall off and Ruby Hill declined. By 1885, only 700 residents remained. Although major mining companies still controlled the claims, activity dropped off. The post office closed on November 30, 1901. A revival took place in 1906 when the Richmond and Eureka companies combined to form the Richmond-Eureka Consolidated Mining Company and built a large smelter at a cost of $200,000. However, a thunderstorm washed out the Eureka-Nevada Railroad in 1910, ending the revival and emptying Ruby Hill. The mines were operated periodically by leaseholders, but no substantial production was ever recorded. The

The Ruby Hill Band in 1912. (Northeastern Nevada Museum)

total production figures for the Ruby Hill District are nevertheless very impressive: Eureka Consolidated, $19.2 million; Richmond Consolidated, $15.2 million; Richmond-Eureka Consolidated, $4 million; K & K Consolidated, $2 million; Ruby Dunderberg Mining Company, $1.6 million; and Jackson Consolidated Mine, $1.1 million. Today, exploration is taking place on Ruby Hill, and plans are in the works to begin production once again. Several buildings still remain in Ruby Hill. Only two, however, predate the 1906 revival. Mining memorabilia and smelter remains are scattered all around and make for interesting exploration. Ruby Hill is a fascinating place to visit, and the short trip from Eureka is well worth the time. (Private property, ask permission)

Safford (Barth) (Pine Mountain) (Pine Valley)

DIRECTIONS: *From Palisade, take poor road south for 4 miles to Safford.*

In August 1881, Ben C. Safford, a prospector, located a rich silver lode. Safford named his find the Onondaga Silver Mine and organized the Safford Mining District. Within a matter of weeks, a small rush to Safford developed. A townsite was laid out by H. N. Fletcher in September 1881. A small camp of about 20 soon formed and buildings were erected, including a schoolhouse. A post office opened on July 31, 1882, with Joseph Tyson as postmaster. He doubled as the town's assayer. Several businesses opened during 1882. Among the more prominent were the Pioneer Saloon (Jonathan Walker, proprietor), the Safford Restaurant (William Tregoning), and the Pratt Board and Lodging House (M. Pratt). An attorney, C. P. Hall, also hung his shingle in Safford. A stage line, from Gerald to Safford, was set up by Jack Petch and ran three times a week.

Although the post office closed on May 7, 1883, the camp of Safford was not declining—yet. Another rich mine, the Zenoli, was discovered by Gabriel Zenoli and Francisco Thoma. Literary enlightenment, the *Safford Express,* arrived on June 2, 1883. The paper, however, was published in Palisade and delivered to Safford because the publisher, Lambert Molinelli, didn't feel that Safford would really last that long. Eventually, his hunch proved correct. Molinelli left the paper in July 1883, and was replaced by W. W. Booth, who had run papers throughout Nevada. The paper, however, lasted only until the end of August before it folded.

Mining continued, with a considerable amount of ore still being removed. An $8,000 shipment of ore was extracted in just one week during 1882. Ben Safford sold his holdings in June 1883 to A. E. Davis, T. H. Kramn, A. Halsey, William Sharp, and L. E. Kelley. These men organized the Onondaga Gold and Silver Mining Company, but the mines ran dry late in the same year, and Safford was soon abandoned. Only one person remained: Ben Safford.

He believed that he would be able to discover a new ore deposit that would bring everyone back to his town. But he died in Safford without ever finding the elusive deposit. The district remained deserted until 1903, when the American Smelting and Refining Company opened the West Iron Mine (a.k.a. Barth Mine) at the mouth of Safford Canyon. During the next fifteen years, the company produced just under $2 million in iron ore. The Zenoli Mine was reopened in 1907 by the Zenoli Silver Copper Company, which produced $67,000 during the next two years. In 1915 the Safford Copper Company, with James Kimball as president, was incorporated and began to mine its Evening Star claim. A test shipment of ore was sent to Garfield, Utah, in 1917, but the ore values were poor and the company soon gave up. After 1918 the Safford Mining District was abandoned for good. The only mining activity in the area took place a few miles north of the townsite. The Nevada Barth Mining Company ran an extensive iron ore mine beginning in 1954. Not much is left in the lonely town of Safford today. Only rubble, mine dumps, and scattered debris remain to recall Ben Safford's dream.

Sherwood

DIRECTIONS: From Bruffey Ranch, continue south for 5 miles to Sherwood. Located 1 mile below Union.

Sherwood was a small offshoot camp of nearby Union. The Sherwood Mine was located here and was active during 1887 and 1888. Only about 15 miners made their homes here. Most chose to live in the more comfortable boardinghouses in Union. Strangely, a post office opened here, rather

The town of Sherwood didn't last very long. (Nevada Historical Society)

than in the larger camp of Union. The office opened on August 9, 1887, with Anna Bowen as postmistress, but closed the following year on July 9. The mines closed after the shallow ore pockets were emptied, and no one has lived here since. Today only mine dumps mark the site.

Sulphur Springs Station

DIRECTIONS: From Romano, head south for 3.5 miles to Sulphur Springs.

Some controversy surrounds the origin of Sulphur Springs Station. While the station is purported to have been a Pony Express stop, no evidence exists that actually shows it was on the original Pony Express line. The station was built in July 1861, in preparation for the opening of the Overland Stage. Since Sulphur Springs Station was located very close to the Pony Express route, it is likely that the riders did occasionally use the spot for horse changes and rest stops before traveling over the mountains to Roberts Creek. The station served the Overland Stage until 1869 when it was abandoned. Today a portion of a log wall and small stone foundations are all that's left to mark the site.

Summit

DIRECTIONS: From Garden Pass, continue north on Nevada 278 for 5 miles to Summit.

Summit was a stop on the Eureka and Palisade and the Eureka-Nevada railroads, at the top of Garden Pass Summit, an altitude of 6,686 feet. The engines took on water here after the tough climb from Eureka. A small camp existed here for many years, first established in 1875 during the railroad's construction. Summit was the end of the railroad for a couple of months until the rails reached Garden Pass. During the 1880s, the camp had a population of about 25. By the turn of the century, however, only a couple of hardy souls still inhabited the camp. The closing of the railroad in 1938 ended Summit's usefulness. Today only scant wooden rubble marks the site.

Tonkin

DIRECTIONS: From Bar, continue north for 10 miles to Tonkin.

Tonkin was a small ranching settlement established just before the turn of the century. The ranch was locally important because of its post office. While Tonkin did not have a rich history, it did have a significant impact on the immediate area. For ranchers in nearby Denay Valley and miners working small mines in the area, Tonkin represented a link to family and friends. The post office opened on December 9, 1898, with John Tonkin as postmaster, and remained open until March 14, 1931. The ranch was taken over by the Damele family in 1906 and is still operating as part of their Dry Creek Ranch. Many of the original buildings remain, including the large ranch house. This is private property, so ask permission before wandering around the site.

Union

DIRECTIONS: Located 1 mile north of Sherwood.

A rich lead-silver deposit was discovered here in early 1879, and a small rush of miners came to the district. By summer, Union had a population of 75 and boasted three stores and two saloons. However, the boom went bust by the end of 1879, and it was not until 1886 that the claims were worked again. James Lindsay gained control of the claims and built a small smelter the following year. Once again the mining activity was short-lived. Lindsay abandoned the district after the mines had produced $100,000, and the area remained empty until 1915.

The Union Mines Company was organized in 1915 by William Fairman of Philadelphia, and Union became a "company town." A two-story bunkhouse and a boardinghouse were built to provide shelter for the miners. The company had four shafts (200 feet, 250 feet, 300 feet, and 550 feet), the two most valuable being the Union and the Armstrong mines. A post office opened at Union on April 27, 1916, with Vina Thatcher as postmistress. The company shipped its ore to Midvale, Utah, where the returns were quite good. However, the cost of the ore removal was extremely high. By the time the company folded in 1918, $225,000 had been produced, but $242,000 had been spent on extracting the ore. When the Union Mines Company folded, Union died. The post office closed on November 27, 1918. While the mines were worked periodically until 1955, the camp never had more than a couple of residents. Today only a few small, crumbling cabins remain at the site. Large amounts of rubble mark the sites of the company buildings. Not much else, except a couple of faint stone ruins, remains in Union.

Vanderbilt (Geddes) (Secret Canyon District)

DIRECTIONS: From Eureka, take U.S. 50 east for 1.5 miles. Exit right onto Secret Canyon Road (very poor). Follow for 4 miles to Vanderbilt.

The town of Vanderbilt formed in 1870 as a result of the many mines located throughout the Secret Canyon area. The camp located near the Vanderbilt Mine, discovered on April 12, 1870. The town grew rapidly, and by the summer there wre 150 residents, three stores, two boardinghouses, and a couple of saloons. More than 300 miners were employed in the district at that time. The major mines surrounding Vanderbilt included the Calico (discovered August 23, 1869), Geddes (August 18, 1869), Stockton, Hodgdon, Page and Corwin, Bayse, Geddes and Bertrand, and Monroe. In December 1870 the Hodgdon Mine was sold to the Sierra Valley Mining and Milling Company for $12,000. It was a good deal for the company. The mine produced enough in the first twenty days to pay back the purchase price. The company constructed a 10-stamp mill that used equipment from the Hot Creek Mill (Nye County). The Sierra Mill cost $40,000 to construct and worked ore from the Hodgdon, Geddes and Bertrand, and Page and Corwin mines.

Because of the numbers of people, a post office opened on August 24, 1871, with Nicholas H. Meating as postmaster. However, nearby Eureka was beginning to boom and many of Vanderbilt's citizens left to go there. Most of the ore mined in the Secret Canyon mines was soon being shipped elsewhere. The Sierra Mill became idle and was sold by the sheriff in 1871 to George W. Chesley of Sacramento for $566. But in June 1872 the mill was sold to Charles Kohn for $7,000. The mill had just been restarted, treating Scorpion Consolidated ore, when a fire totally destroyed it in September 1872.

This disaster just about finished Vanderbilt. The post office closed on August 8, 1873. While sporadic mining activity took place in Secret Canyon during the next few years, most miners lived in Eureka. By the time a small revival began in 1880, only 25 people were left in the district. The main mining company in the district during the 1870s and 1880s was the Geddes and Bertrand Mining and Milling Company, which had been incorporated in 1872. The company controlled the Geddes and Bertrand Group and, despite the mining slowdown during the middle and late 1870s, continued to produce into the 1880s. The company built a small, but very expensive, mill and furnace with a 20-ton capacity in 1880 at a cost of $300,000. Because of the revived mining activities, some people returned to the camp, and on March 17, 1882, the post office reopened. The post office was renamed Geddes, after Sam Geddes, president of the Geddes and Bertrand Company. While the post office closed on June 18, 1885, Geddes and Bertrand continued to be active until 1896, although its involvement sharply declined after 1887. A major financial blow to the company occurred on August 15, 1886, when a fire destroyed the mill and furnace. While several mines were leased off and on during the next forty years, no substantial production was recorded. The total production for the district from 1867 to 1940 was $722,000. The bulk of this production, $630,000, was by Geddes and Bertrand. Today only scant mill ruins mark the site, and because the road to Vanderbilt is extremely treacherous, it is not worth the risk to try to reach the site.

White (Bailey Ranch)

DIRECTIONS: *From Romano, continue north for 2.5 miles to White.*

White was a small ranching settlement formed by James White in the 1880s. While not more than 15 people ever lived at White, the settlement was important in the lives of the people living in Diamond Valley. A post office opened here on July 31, 1890, with White acting as postmaster. This allowed ranchers and miners in the valley the luxury of not having to travel to Eureka to get their mail. In 1899 Robert Bailey bought the ranch, but he did not want to continue the post office. The office at White closed and was reopened two years later in nearby Romano. The ranch continued in operation until the 1960s, but it is abandoned today. The remains at White are quite extensive and show the style of a turn-of-the-century ranching operation. A unique relic at the site is an old wooden slaughter wheel, an item rarely seen today.

Willards

DIRECTIONS: *From Hay Ranch, continue north on Nevada 278 for 4 miles to Willards.*

Willards was a stop on the Eureka and Palisade and the Eureka-Nevada railroads. The stop was named for the Willard family, who ran a ranch next to the railroad. The main ranch house, which served as a boarding-house and restaurant for railroad passengers, burned in 1916. Although no real settlement developed here, a consistent population of 25 was reported for many years. Today the Willard ranch is still active. Many of the buildings date from before the turn of the century, and one of the more prominent structures was associated with the railroad, as either a small depot or a freight building.

A Short History of Lander County

Lander County, the largest in the state, was created on December 19, 1862, soon after discoveries were made at Austin. White Pine County was carved out of Lander on April 1, 1869, Elko County on March 5, 1873, and Eureka County on March 1, 1873.

The "Rush to Reese River" during the early 1860s led to a huge influx of people. Austin and many smaller towns sprang up, but most faded just as quickly. The Austin discoveries were the first major strikes in Nevada since the Comstock Lode. Prospectors combed the mountains and canyons throughout Lander County. Towns such as Kingston, Canyon City, Yankee Blade, Clinton, Geneva, and Amador flourished briefly.

The Pony Express played an important role in Lander County's development. A number of stations were established in the county, and one of them, Jacobsville, became the first county seat. It was a Pony Express employee from Jacobsville who made the first Austin discoveries.

The completion of the Nevada-Central Railroad in 1880 opened the heart of Lander County to easy rail access for mining and ranching interests. But mining slowed dramatically until after the turn of the century. Then Tenabo, Hilltop, and Betty O'Neal boomed through the 1920s. During the next forty years, however, only Galena had significant success.

Microscopic gold has become the new boom for Lander County. Mines near Battle Mountain, Big Creek, Buffalo Valley, Gold Acres, McCoy, and Cortez have already produced more than was mined in the county during the first hundred years. Production from 1862 to 1969 was more than $110 million. During the last thirteen years, that figure has doubled.

Battle Mountain now serves as the county seat. Austin, although it lost the seat in the 1980s, retains its quaint historic flavor. Lander County has had an illustrious history and with renewed mining activity, continued successful ranching, and a growing tourist trade, the county's future is bright indeed.

Lander County

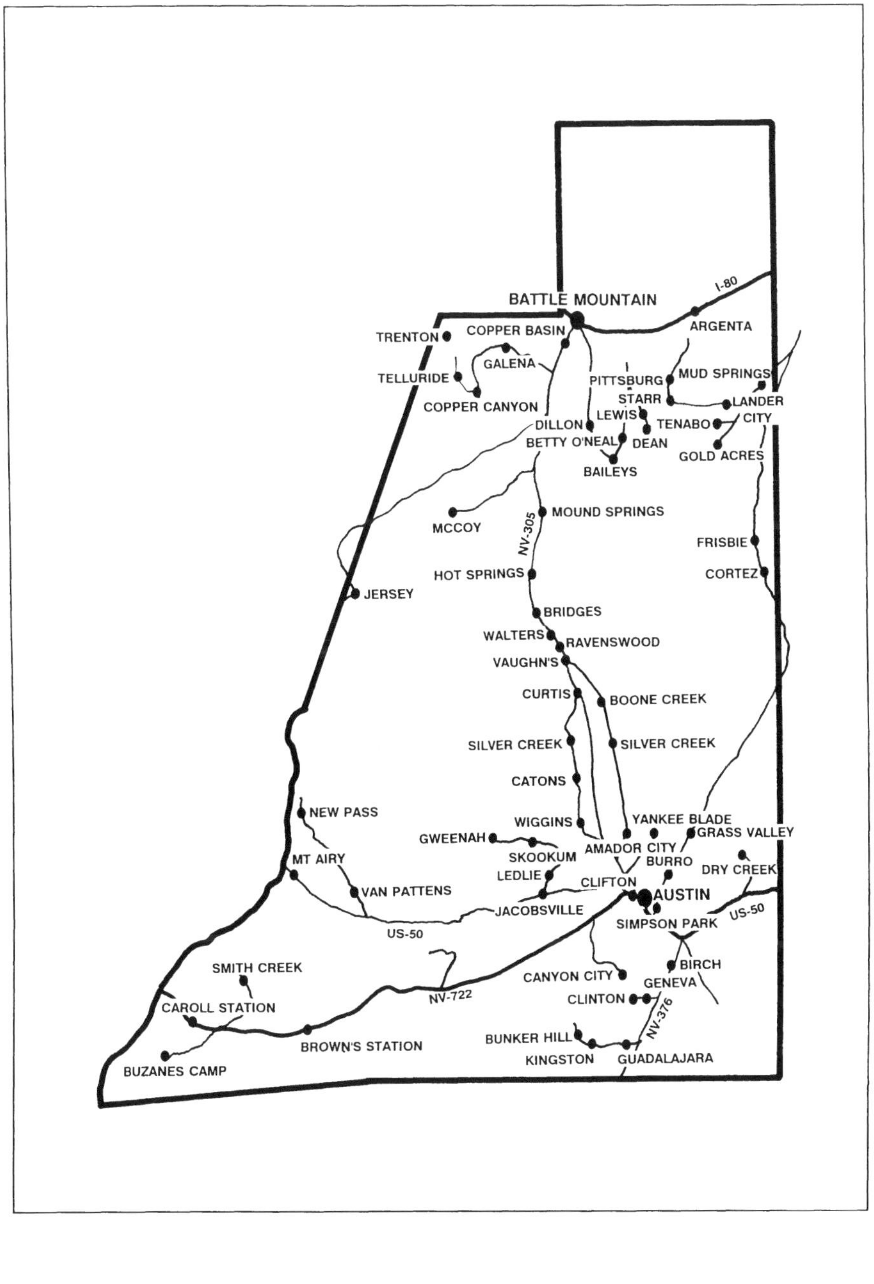

BATTLE MOUNTAIN
I-80
ARGENTA
TRENTON
COPPER BASIN
GALENA
TELLURIDE
COPPER CANYON
PITTSBURG
MUD SPRINGS
STARR
LANDER CITY
LEWIS
TENABO
DILLON
BETTY O'NEAL
DEAN
GOLD ACRES
BAILEYS
MOUND SPRINGS
MCCOY
NV-305
FRISBIE
HOT SPRINGS
CORTEZ
JERSEY
BRIDGES
WALTERS
RAVENSWOOD
VAUGHN'S
CURTIS
BOONE CREEK
SILVER CREEK
SILVER CREEK
CATONS
NEW PASS
WIGGINS
YANKEE BLADE
GWEENAH
GRASS VALLEY
SKOOKUM
AMADOR CITY
MT AIRY
BURRO
DRY CREEK
LEDLIE
CLIFTON
VAN PATTENS
AUSTIN
JACOBSVILLE
US-50
SIMPSON PARK
US-50
SMITH CREEK
BIRCH
CANYON CITY
GENEVA
NV-722
CLINTON
CAROLL STATION
NV-376
BUNKER HILL
BROWN'S STATION
KINGSTON
GUADALAJARA
BUZANES CAMP

Amador (Coral City)

DIRECTIONS: From Austin, head west on U.S. 50 for 0.6 miles. Exit right on Nevada 305 for 2.3 miles. Exit right and follow for 1.5 miles to Amador Canyon. Then park and follow footpath to right for 1.5 miles.

After the rich discoveries throughout the Reese River District, prospectors combed every canyon in the area looking for new ore deposits. During the spring of 1863, silver ore was discovered here, and a rush to Amador Canyon resulted. A townsite was laid out in the summer by people named B. T. Hunt, Bowe, Chase, Matheny, Meek, and Kinsey. The mining district was officially organized in November, and by the end of the year about 200 hardy souls were clustered on the side of the cold mountain. As the camp continued to expand, a smaller adjacent camp, Coral City, was absorbed into Amador. A post office, with Isaac Sherman as postmaster, opened on April 16, 1864, giving Amador an air of permanence.

The original discovery at Amador was the Coral Mine. By 1864 several mining companies were working the district: Mammoth Tunnel and Water Company, Monte Christo Mining, Wildwood Tunnel and Water, Aspinwal Gold and Silver Mining, Amador Consolidated Silver Mining, Silver Age Mining, Dashing Wave Gold and Silver Mining, and Loomis Gold and Silver Mining. Many of the ore deposits around Amador occurred in valuable ledges, the best of which were the Aspinwal and the Bigelow ledges, owned by Peter Davis. Considered by many to be the finest in the Reese River District, these ledges were assayed in January 1864 at $295 and $162 per ton, respectively. Another valuable ledge was the O'Hara, which was sold in August 1864 for $32,000.

Yet permanence was not in the cards for Amador. The ore deposits were not as extensive as they were first thought to be, and most of the mines began running empty by early 1866. The boom was over and Amador faded quickly. The post office closed on April 24, 1866, and Amador's businesses began to leave the camp. By the end of that year, the camp was virtually abandoned. The camp died a quick death and was never an active producer again. By 1869 the last resident had left, consigning Amador to the ghosts. A trip to picturesque Amador Canyon has many interesting stone ruins to offer. As you arrive in Amador, you can experience the sense of isolation that the residents of these small camps felt. Although the walk via trail is difficult, consider making the trek. The view itself is worth the effort.

Argenta (Battle Mountain Station)

DIRECTIONS: *From Battle Mountain, head east on U.S. 80 for 12 miles to Argenta, located on the north side of the railroad tracks. The Argenta Mine is located 2 miles to the southeast.*

Silver was discovered in 1866 at Argenta, and a small camp quickly formed. Argenta became the first Central Pacific Railroad stop in Lander County for passengers heading east. A post office, with William Westerfield as postmaster, opened soon after on December 4, 1868. The town became a shipping point for the Austin mines, and residents had high hopes that Austin would help make Argenta the railroad center of Lander County. However, attention soon began to focus on nearby Battle Mountain, which was not only closer to Austin but nearer to the booming Galena Mines as well. Argenta's residents realized that the town was doomed, and in December 1870 they moved everything—buildings and themselves—to Battle Mountain. By 1873 only the railroad signal station and a few buildings remained at Argenta. The post office finally closed on February 24, 1874, and Argenta became a ghost town. It was not until 1930 that interest in Argenta was revived, with the discovery of rich barite deposits on Argenta Mountain. The Argenta, or Barium King, Mine was established and is still producing today. From 1930 to 1969, well over $3 million in barite was produced. Today, a mill is running at the Argenta siding, where barite is loaded directly onto the trains. Several cattle shipping corrals are also at Argenta. But nothing from the early years remains.

The Argenta railroad depot in 1869. (Nevada Historical Society)

Austin

DIRECTIONS: Located on U.S. 50.

On May 2, 1862, William Talcott, a former Jacobsville Pony Express agent, journeyed to Pony Canyon in central Nevada Territory. Pony Express riders, who used it as a shortcut from Simpson Park, had named it in 1860. There, Talcott found an outcropping of silver ore, which he called the Pony Ledge. Rich assay reports came back, and the mining district was organized on May 10 and established on July 17. The boom was on.

A small town formed as the rush to the new district developed. Clifton, located just below the original claims, was platted by two people named Marshall and Cole. By spring 1863 a tent city of 500 had sprung up at the mouth of Pony Canyon. Flora Bender, while traveling through Clifton, reported, "The houses are principally of canvas, with roofs made of cedar brushes but very few wooden buildings."

Clifton's businesses included Wells-Fargo, assay and recorder offices, hotels, restaurants, and lumber and hay yards. A justice of the peace was elected. As Clifton was becoming a prosperous town, a new camp blossomed a mile farther up Pony Canyon.

Austin's townsite was laid out by David Buel. He named it Austin for his partner, Alvah C. Austin. Many differing accounts about the naming of Austin exist. Other people mentioned as possibilities for whom the town could have been named were John Austin (pioneer), George Austin (developer of the Jumbo Mine in Humboldt County), and Leander K. Austin (uncle of George). Others believed the site was named for Buel's hometown, Austin, Texas.

The camp grew quickly. Lander County was formed on December 19, 1862. At the time, the county constituted almost one-third of Nevada. By 1863, almost 10,000 people had flocked to the Austin area. Austin's boom prompted the territory's legislature to move the county seat from Jacobsville and to the young silver mining camp. Lots on main street sold for an average of $8,000 in gold.

Although a post office did not open until November 20, 1863, the bustling town was already becoming a mining center. By May, Austin had two hotels, five saloons, a telegraph office, and a registered voting population of 450.

A mining milestone occurred on May 28, 1863. The first silver brick was produced from ore crushed in a Mexican *arrastra* on the Tesora claim. An *arrastra* was a crude ore grinder that used horses to pull heavy stones across the ore.

In August, Buel and his company completed the first stamp mill. The initial ore treated by the mill was a 10-ton shipment from the Morgan and Muncy Mine that yielded $3,360.

The first newspaper arrived in 1863 in the form of the *Reese River Reveille*, owned by W. C. Phillips. The newspaper was destined to become an enduring

Austin in 1890. (Special Collections, Library, University of Nevada, Reno)

publication. It is still being printed today and claims to be the oldest continuously published newspaper in Nevada. A six-column, four-page, weekly journal, the *Reveille* was staunchly Republican, as was its owner.

A school opened on July 22, 1863, in a pine bough shelter. The International Hotel, a landmark of Austin, was first built in Virginia City in 1860 and late in that decade was moved 180 miles east and reconstructed on Austin's main street. The Overland Telegraph, which began operations in spring 1863, enjoyed a booming business, with more than $2,500 in receipts for October alone. The office became the third most profitable in the West Coast area. By the time of Austin's incorporation in 1864, the town's voting population had grown to 6,000.

Austin and Clifton became embroiled in intense competition for supremacy. Clifton had the advantage of a level townsite. Austin began offering free lots for businesses in exchange for help in building a graded road from Reese River Valley to Austin. The new road to Austin bypassed Clifton. Although Clifton was slightly larger, it had lost the battle. The combination of Clifton's expensive lots and Austin's free lots spelled doom for Clifton.

When Austin won the contest for the county seat in 1863, people began to move to Austin. Finally, on February 20, 1864, the Clifton post office closed. Soon Clifton was left with only a few empty wooden buildings, small islands in what had once been a sea of tents. A small 4-stamp mill, the Clifton, was still active in 1865, but interest in Clifton was gone. When the mill closed in 1867, the town was abandoned.

But the post office, for some unknown reason, was reopened, and Frederick May served as postmaster from June 20, 1867, to September 28, 1868. It was clear, though, that even the postal system had little interest in Clifton. Austin was growing by leaps and bounds, and the unfortunate town of Clifton was left in the dust.

By 1864 the district had become a force in Nevada's mining production. Total production through January 1864, was $100,000, and that was the lowest annual production for many years. Mills were built throughout the district. In early 1864 the Isabella Company erected an experimental roasting furnace. By August 1865, nine stamp mills were running in Austin.

Stamp mills were so named because of heavy stamps, or pulverizers, used to crush ore. New mills included the Oregon (renamed Manhattan, 20 stamps), Long Island (5 stamps), Hildreth (5 stamps), Union (10 stamps), Ware (5 stamps), Boston (20 stamps), Eagle (10 stamps), Pioneer (10 stamps), and California (10 stamps). Several others were also active in the mining camps of nearby Big Creek and Yankee Blade. In 1866, the Silver Hill, with 5 stamps, was built. The Manhattan was expanded to 30 stamps and the Pioneer to 20.

For a while, during the 1860s, some companies used camels for hauling. Sam McLeneghan had originally used a few camels to carry freight from Sacramento to towns in Nevada. The idea was unsuccessful, and he sold them in Austin. The camels hauled salt across the desert from Walker Lake. More efficient methods of transportation were soon employed, and the camels were no longer useful.

The major mining firm in Austin was the Manhattan Silver Mining Company. The company bought out the Oregon Milling and Mining Company in June 1865. Between 1865 and 1887, Manhattan produced half of the district's silver bullion. The company also gained control of the Manhattan mill. When the mill was expanded to 30 stamps, fourteen amalgamating pans, seven settling tanks, and a furnace were added. This expansion was interesting. The 10 new stamps were located not in the mill but in an adjacent building. The mill employed 50 men.

Despite the expansion, there was no smooth sailing for the company. Early financial troubles forced seizure of company property by the sheriff in 1866. The problems were resolved and activity resumed later that year. The original property consisted of several mines and claims, including the Southern Light,

North Star, Blue Ledge, and Oregon. The mines had workings of more than 3,800 feet and produced more than $67,000 in 1865. This promising start was only a sign of future production. In 1867, the North Star alone produced $252,000, with its sister mines boosting the total to $339,000. Lack of labor forced the company to curtail operations temporarily in 1869 when the White Pine rush drained Austin of miners.

Austin's production continued to increase. From July 1866 to August 1867 its mines produced more than $1.4 million. Mills continued to play an important role. The Citizen's Mill, built with machinery removed from the Knickerbocker in Nye County, joined the mill force in 1871. The mill was unsuccessful and closed in May 1872. Union Mill was moved to Hot Creek, Nye County, in August 1867.

Manhattan Silver Mining was by no means the only company active in Austin. Within three years of the first discoveries, more than 600 mining companies were formed. Although most of these companies folded by 1870, some survived. Two of the largest were California Mining and Milling, and Pacific Mining. Along with the Manhattan company, these firms controlled most of the valuable mines.

The richest mines were Panamint, with total production of $9 million; Paxton, $1 million; Buel–North Star, $2 million; London, $1.5 million; Independence, $3 million; Oregon, $5 million; Isabella, $3 million; Union-Whitlatch, $3.8 million; Savage-Diana, $2 million; Silver Chamber, $1.5 million; and Lane and Fuller, $325,000 in three years. Of these mines, the Oregon was the deepest and the most extensive.

While mining activity was at a peak during the late 1860s and early 1870s, production began to slow down rapidly after that. In 1872, as mining was ebbing, union activity appeared in town. In February a miners' union was organized. Although a majority declined, a small number of miners did join. Shortly after organizing, a group of 40 union men marched on the Lane and Fuller Mine and demanded $4 a day in wages. Twenty-nine miners were arrested. Despite this show of strength, the union was not effective because of the multitude of small properties and leaseholders in the district. During this period, Manhattan Silver Mining and Pacific Mining employed 170 men. Small property owners and leaseholders outnumbered company employees.

The last robust year for Austin was 1872, with a production of $250,000. Pacific Mining was forced into bankruptcy in August 1873. Sixty-eight claims became idle. Manhattan was left as the only viable mining company in the district, and many minor companies merged with the financially secure Manhattan. The company was easily the primary producer from 1873 until it folded in 1887. Austin was not dying—it was just settling down to a steady existence.

Austin had been incorporated on February 17, 1864. Morris Locke designed a city seal that commemorated the famous Gridley Sack of Flour.

The Gridley store in 1985, before restoration.

The story goes like this: Reuel Colt Gridley, who operated a general merchandise store in Austin, made a wager on the mayoral election in 1864. Because he lost the bet, he had to carry a 50-pound sack of flour from Clifton to Austin. The sack was decorated with red, white, and blue ribbons and numerous flags. The band played "John Brown's Body" while Gridley marched. Then the sack of flour was auctioned off to raise money for the Sanitary Fund, a forerunner of the American Red Cross. After someone bought the sack of flour, it was returned to be sold again and again. More than $6,000 was raised that first day. Gridley and his famous sack of flour traveled throughout Nevada and California, eventually raising more than $250,000. Gridley later moved to California, where he opened a grocery store in Paradise. He died on November 24, 1870.

The Gridley Sack of Flour wasn't the only fund-raising idea in Austin. In 1863 a committee was formed to raise money for a school building, but donations were slow. David Buel had a unique pair of shoes, very large with high buttons. The shoes were "stolen" and auctioned off in downtown Austin. Every time the shoes sold, they were returned to be resold. The Austin Buel Shoe Fund was established. By time Buel got his shoes back, more than $100 had been raised.

Newspapers found fertile ground in Austin. The most prestigious was the *Reese River Reveille,* which began publication on May 16, 1863, selling at a price of 50 cents. Owner W. C. Phillips started with a semiweekly, but in response to demand, he enlarged the paper to a triweekly in November 1863. Myron Angel was the editor of the *Reveille* from February 1864 to January 1868. Angel later worked on the famous 1881 book *History of Nevada,* published by Thompson and West. In May 1864, Phillips fell ill and sold the *Reveille* to Oscar and Jacob Fairchild. Phillips returned to his home in Illinois, where he died six months later. Almost immediately, the Fairchilds made the newspaper a daily with a new name, the *Daily Reese River Reveille*. The paper remained a daily throughout Austin's boom days before becoming a weekly on September 11, 1869.

On August 14, 1871, the Fairchilds abandoned the newspaper business. The paper was sold to Andrew Casamayou and John H. Dennis. They were well liked by Austinites, and the paper prospered. By 1873 Dennis had had enough of newspapers and sold his share to John Booth. Casamayou took over the editorial duties. Booth and Company took complete control on December 21, 1875, when Casamayou died suddenly. The two men had been simultaneously running the *Reveille* and the *Belmont Courier*. It is likely that the strain of running two papers led to Casamayou's untimely death.

During the 1870s, one of the most popular sections of the *Reveille* was initiated when editor Fred Hunt started the famous—or infamous—Sazerac Lying Club, which soon was a regular feature. "Sazerac" was the name of a local saloon and a popular brandy of the time. People tried to outdo each other with outlandish tales, which were duly recorded by Hunt in the *Reveille*. Many big-city newspapers used the stories, and a strong following developed.

Booth continued to run the *Reveille* until his death in March 1884. Then his widow ran the paper until she sold it to George Carpenter in January 1889. Carpenter struggled for a while, then sold to W. D. Jones in 1893. Jones changed the paper to a semiweekly and brought the *Reveille* into the twentieth century.

There were other newspapers in town as well. The *Austin Republican* (1868), *Peoples Advocate* (December 1890 to January 1893) and the *Daily Morning Democrat* (August 1882 to July 1883) also were published. The most important of these was the *Democrat,* run by W. W. Booth, son of John Booth. The paper served as a Democratic party mouthpiece, which put it at odds with the heavily Republican *Reveille*. The *Democrat,* costing $12 per year, delighted in taking jabs not only at the *Reveille* also but at other Republican newspapers throughout the state. The *Democrat* proudly proclaimed itself the "only reliable, readable, newsy, and largely circulated daily published in Lander County."

The paper was one of the few that demonstrated an active sense of humor. Once the *Tuscarora Times-Review* used an article from the *Democrat* but didn't give the paper a credit line. The *Democrat,* in its next issue, declared: "We presume we must excuse our contemporary, as he is likely so imbued with Republican principles, as to be affected with kleptomania."

The famous Mules' Relief, which brought people and supplies up the steep grade from Clifton to downtown Austin. (Nevada State Museum)

Despite the quality of the paper and the loyal following it enjoyed, Booth decided to move on and start papers at other towns, including Belmont. Finally he settled in Tonopah to run the *Tonopah Times-Bonanza*.

After the short tenure of the *Peoples Advocate* in the early 1890s ended, the *Reveille* was the only literary source in Austin.

Mining around Austin slowed, and by 1878, production reached only $163,000. Austin continued to be an important town, however. The Nevada Central Railroad was completed to Clifton on February 9, 1880. Austinites wanted the railroad to run on into town, and construction began in May 1880 on the Austin City Railway. The 2.8-mile, three-foot narrow gauge extended from Clifton, through Austin, and ended at Manhattan Mill. Clifton, now known as Austin Junction, was revived to a small extent when it became the terminus for both railroads. A 30-car siding for the Nevada Central was built. This was the only purpose of the old town and nobody moved back to it.

By August, work had progressed on the Austin City Railway to the International Hotel on Main Street. It wasn't until May 31, 1881, however, that the first trial run was made, and another month passed before actual work trips began. The engine was christened the *Mules' Relief,* since mules no longer had to pull heavy loads up the steep canyon from Clifton.

While successful, the line was not without tragedy. Around 5:30 p.m. on August 19, 1882, the *Mules' Relief* was heading down to Clifton when control was lost. Three men were aboard: brakeman William Reagan, fireman Frank Duffy, and engineer Andy Wright. The engine jumped the tracks at a sharp corner halfway down the canyon. Reagan and Duffy jumped to safety, but Wright was trapped under the boiler and killed instantly. A moving memorial to Wright appeared in the *Democrat* the following day:

Honor to the Brave

In this little mountain town;
Deeds that should have gained renown,
And have won the Hero's Crown,
Have oft been nobly done
Ignored in song and story
No less these heros' [*sic*] glory,
Than those of conflicts gory,
Of battles fought and won.

Through burning shafts descending
Miners have gone depending
On one frail rope intending
A comrade's life to save.
When such gallant deeds are done
In history they should run,
With those who have fought and won,
Their fights or field and won.

Where our mountains have been rent
By fierce volcano's vent,
A canyon's steep ascent
Leads to our town
While upon the canyon's side
The rails are spiked and filed
And make steepest railroad ride
Where mountains frown.

In snake-like curves the rails creep
Along this vast canyon deep
And climb to summit steep
Where lofty mountains soar
In less than one short mile's space
The engine in giant race
Ascends the mountain's face
One thousand feet or more.

Then descending from this height
Taken by momentum's might,
The engine resumed her flight,
And like a meteor sped.
When in the town, her light was near,
And in the gloom would disappear
Pray God, help her brave engineer
We all thought, if not said.

The engine to the valley sped
The brake applied but still she fled
Control is Lost brave Andy said
To his fireman bold,
Down the canyon, headlight flashing
Down, over the steep, headlong dashing
Around the curves swiftly crashing
The rest is soon told.

Cool, unflinching, tho' death was near,
At the lever stood the engineer
Trying his best to reverse the car
But speed was at its might
While at the lever he tried
Jump Frank, to his fireman he cried
And it was thus our hero died
Engineer Andy Wright.

Thankfully, this tragedy was an isolated incident. The *Mules' Relief* was soon repaired and operated without mishap until the line closed in 1889.

Natural disasters made occasional appearances in Austin. Cloudbursts in Pony Canyon would funnel into town, turning the main street into a river. After minor floods in 1868 and 1869, Austin was hit with a major washout on August 13, 1878. Though damage estimates were sketchy, it is clear that many Main Street buildings were destroyed. Luckily there was no loss of life. Other floods caused extensive damage in 1874, 1884, 1891, and 1901. With almost every good rain, Austin became a quagmire of mud.

Floods weren't the only unwelcome visitors to Austin. Several fires damaged parts of town. The largest occurred on August 9, 1881, destroying one side of the main street. But the buildings were rebuilt almost before the ashes cooled.

By 1880, Austin's population was 1,700. With the completion of the Nevada Central Railroad, other railroads were planned. The Nevada Southern Railway, organized in November 1879, was proposed to run from Austin through Cloverdale and Ione Valley in Nye County. From there the line was to run on

Main Street, Austin, showing the effects of the August 15, 1878, flood. (Nevada Historical Society)

to Columbus and Silver Peak and eventually join the Southern Pacific. J. R. Hudson surveyed the route in March 1880, but completion of the Carson and Colorado Railroad ended hopes for the line from Austin. The new link made a railroad from Austin superfluous, and the plans were canceled during the summer of 1880.

The Union Pacific also planned an extension from Austin to Silver Peak in 1881. Union Pacific bought the Nevada Central in June 1881 for $450,000 and immediately began surveying a route to Silver Peak. But Union Pacific soon realized it was an unprofitable venture and dropped the plans. The company lost interest in Nevada Central, and the line went into receivership in October 1884. The original owner, Anson Stokes, bought the railroad at a bankruptcy sale.

In November 1888, the reorganized Nevada Central Railroad began operations. In 1902, another railroad, the Nevada Midland, was planned, to run to Tonopah, but the idea soon fell through. The Nevada Central people surveyed their own route to Tonopah in 1905 but soon realized that a project of that size was unfeasible and stopped construction. The Nevada Central continued to run uneventfully until it was abandoned on January 31, 1938. After 32 years of profit and 25 years of losses, the railroad was no more.

Mining in Austin came upon hard times during the 1880s. The Manhattan Silver Mining Company continued to be the only major producer in the district. The company was forced to fold in 1887 after producing $8.6 million. Most of the production, however, came before 1882. The best year was

1881, when $660,000 worth of ore was mined. After Manhattan's mill closed in August 1887, Austin's mines were inactive for many years. The Manhattan property was acquired in December by the newly formed Manhattan Mining and Reduction Company.

In 1888 the company began producing ore from the Plymouth and Union mines. By 1890, Plymouth had been abandoned, and mining concentrated on Union, which employed 45 men. Manhattan produced $287,000 from 1888 to 1891, but the cost of producing the ore was $272,000.

Austin's population was still more than 1,200 in 1890, and the mining industry continued to struggle. In September 1891, Austin Mining Company bought the Manhattan property and deepened the Union Mine from 200 to 700 feet. In 1894, J. Phelps Stokes, president, decided to construct a 40-stamp mill, the Clifton, at the mouth of the Austin-Manhattan Tunnel. Clifton Mill was equipped with machinery from the old Manhattan Mill and was in operation for more than a decade.

However, during that time Clifton never had any permanent residents. The major project of the company was the Austin-Manhattan drainage tunnel. Between 1891 and 1901, the tunnel was extended 6,000 feet into Lander Hill. The tunnel tapped into the lower sections of the Lander Hill mines in 1896, helping to drain excess water that was hampering operations. The concept was similar to that of the better-known Sutro Tunnel that drained the Comstock Lode.

Austin Mining was dealt a major financial blow in 1898 when the manager embezzled $300,000. The company had produced $482,000 between 1891 and 1898, but the loss of $300,000 was too much to overcome, and Austin Mining folded in 1901, shortly after the tunnel was completed.

There was a lull before activity resumed in 1905, and the lack of mining resulted in a decline in Austin's population to 700. By 1905, the principal mines were controlled by Austin Hanopah Mining Company. Very little mining was actually done. Activity finally picked up in 1907 when the Nevada Equity Mines Company began to work some claims. The company was rewarded with a rich strike in the Jackpot Mine. In 1908, Austin Hanopah was taken over by Austin-Manhattan Consolidated Mining.

Work was immediately begun on the Clifton Tunnel and the Union Mine. Austin-Manhattan Consolidated controlled most of the same mines that had been owned by Manhattan Silver Mining. The company tried to do too much work for the available money and was forced into receivership in 1911.

The Jackpot Mine provided the lone bright spot for Austin mining. The mine produced $80,000 between November 1910 and August 1911. Its success created renewed interest in Austin, and new companies moved into the district.

In 1914, the Austin-Dakota Development Company began work on the OK Mine. The OK Mine produced more than $15,000 in 1918. The Jackpot Mining Company took over the Jackpot Mine. During the teens, Maricopa Mines Company became active. The company controlled 33 claims and owned a 100-ton concentrator in New York Canyon.

Nevada Gold Mines Company, with J. S. Madden as manager, worked several mines, including Gold Park, La Crux, May Do-So, Cottonwood, and San Pedro.

Austin-Nevada Consolidated Mines, with H. G. Richardson as president, also began operations. This company, which owned the Austin-Nevada group on Lander Hill, became the most prominent of the new ventures. The company gained control of the X-Ray group in Marshall Canyon, which averaged $300 per ton in ore removed, but it wasn't enough and the company stopped mining in 1921.

Despite a few bright spots, Austin's production figures weren't that impressive. From 1902 to 1921, Austin mines produced only $265,000. Most mines were completely idle during the 1920s. Only a couple of small companies tried to scratch out a living from the old mine dumps. Despite this lack of activity, almost 600 people still lived in Austin.

Austin Silver Mining Company worked the Isabella and Magnolia mines and reprocessed the X-Ray, OK, and Pony mines dumps. The company controlled 150 claims, including 5 in the nearby ghost town of Yankee Blade. Active from 1921 to 1938, Austin Silver Mining produced only $78,000.

The last mining companies to prove productive in Austin were the Kilborn Equity Group and Castle Mountain Mining Company, which operated near Stokes Castle and constructed a small mill. This activity lasted only from 1947 to 1950, and after these companies folded, mining activity around Austin was essentially ended. Total production for the Austin mines has been estimated anywhere from $50 to $65 million.

Today, with a resurgence of mining, there has been renewed interest in Austin. A small mill was recently built on the outskirts of town. A new open-pit operation at Big Creek, just south of Austin, has given the town new life, and a local turquoise mine has had success during the past decade.

Austin had been the county seat of Lander County almost from the start, but in the late 1940s pressure began to grow to change the county seat to the larger town of Battle Mountain. Austin's population had shrunk measurably, and now stood at only about 200. During 1953, the battle became heated. On May 30, 1953, the *Reveille* reported:

> A story is being circulated in Battle Mountain to the effect that some Battle Mountain woman, unidentified of course, narrowly escaped injury when a large portion of the plaster ceiling in one of the courtroom offices fell down and narrowly missed crushing her. As every ceiling in the courthouse is of matched board, not plaster, there seems something wrong with that story. While it may have been possible that the

Battle Mountain woman got plastered in Austin, she certainly didn't get plastered in the courthouse.

While the pressure was constant, it was not until 1980 that Austin lost the struggle and the county seat was moved to Battle Mountain.

Austin has been home to some prominent Nevadans. The foremost among them was Emma Wixom, better known as Emma Nevada. Born in 1859 at Nevada City, California, she moved to Austin soon after, where her father was a physician. Wixom became famous as an opera singer and was nicknamed the Comstock Nightingale. She even made an appearance in Austin with her troupe for the benefit of the Methodist church. Until her death in 1940, she was a source of civic pride for the town.

Clara Dunham Crowell became the first woman sheriff in Nevada when she took over the post in 1919 after her husband, George, died. She achieved a reputation as a tough law officer and was highly respected by the townspeople. She retired in 1921 to work as an administrator of the county until her death in 1942.

Another resident, Anson Stokes, made his own mark on Austin's history as the builder of the famous Stokes Castle. This structure, which still stands, is one of Austin's most memorable landmarks. The structure was started in the fall of 1896 and completed in June 1897. Stokes built the building for his Yale graduate son, J. G. Phelps Stokes. The Tower, as it is locally known, was used as a residence, but it served another purpose as well. The Stokeses used the building to stop high-grading of ore from Stokes Mine, located just below the Tower. By spying on exiting miners, they were able to pick out those stealing ore and catch them before they left the property. After 1897, the structure was unoccupied. A member of a very prominent Eastern family, Anson Stokes was well known as a mine developer and railroad magnate. He controlled many of Austin's mining interests during the town's most productive years.

For many years Austinites used Clifton as a recreational area. A baseball diamond was built and Austin had a competitive baseball team. A small park was also developed and used as a picnic area. Both are in complete disrepair today.

Most buildings were eventually moved to Austin or torn down for lumber. The mill ruins dominate the site, under the watchful eye of the Stokes Tower. The Austin Cemetery, located just below Clifton, offers information about a cross-section of Austin's culture and people.

Today Austin is a quiet town of about 200 people and remains a historical masterpiece. Many of the original buildings have survived and help the town retain an intriguing vintage flavor. Several churches remain, and the varied architecture of the different denominations is fascinating. The number of diversified religions represented is unique to Nevada's pioneer towns.

One of Austin's many old landmarks.

The Gridley store, a stately stone structure at the east end of town, has been recently restored, and a state historical marker is located there. The International Hotel has continued to operate and still caters to tourists and locals. Other historic buildings, the Masonic and Odd Fellows halls, and the Austin City Railway engine house remain. The brick county courthouse still stands proudly, the stubborn centerpiece of a town that refuses to die.

What makes Austin exceptional among Nevada towns is that relatively few new buildings have been constructed. As a result, there has been little destruction of historic buildings to make way for new structures. Austin offers so much that it will take a couple of days to see and appreciate it all. A walk along Austin's streets transports one back to a bygone era and provides the historian and the casual tourist with a rare glimpse of what life was like 120 years ago.

Bailey (Baily)

DIRECTIONS: *From Dillon, continue south for 2 miles to Bailey.*

Bailey was a three-car siding on the Nevada Central Railroad starting in 1880. While the railroad was being built, Bailey was used as a base camp for workers. Enough people lived here that on January 5, 1880, a post office opened, with Edward Bailey as postmaster. The office remained active until November 12, 1887. Bailey soon faded into obscurity, with its only mention in history being on February 13, 1910, when five miles of track were washed out nearby. Nothing remains at the desolate site.

Bannock (Limelite)

DIRECTIONS: From Battle Mountain, head south on Nevada 305 for 12 miles. Then exit right again and follow for 2 miles to Bannock.

Alex Walker and Sherman Wilhelm discovered a rich gold ledge in Philadelphia Canyon during the summer of 1909. A small rush to the area developed that August. Walker named the district Limelite and a townsite was soon laid out. Among the more prosperous mines in the district were the Pussin' Ken (original ore assayed at $180,000 per ton, first 300 pounds brought $2,400), Reno ($200 per ton), and Washoe ($80 per ton). Soon after Walker and Wilhelm made their discoveries, valuable placer deposits were also discovered, adding to the excitement in the district. Wilhelm bought Walker's Limelite Mine and the adjoining claims for $22,000 and formed the Nevada Omaha Mining and Milling Company. One of the investors in the company was A. C. Mohler, vice president of the Union Pacific Railroad. Wilhelm had a lot of problems with claim jumpers. Once a man kicked down claim markers on his property, and after the man was caught, Wilhelm forced him to strip to his underwear and then sent him on his way!

The boomtown of Bannock, shortly after its founding.

By the time a post office opened on November 5, with Parker Liddell as postmaster, quite a few businesses had become established. Bannock (named for the Bannock Indians) had a population of almost 200. The first business to open was the Mayer Saloon, run by C. A. Mayer. Mayer also owned the Philadelphia Club and co-owned the Mayer-Reber Mining, Leasing, and Development Company, which bought and sold mines and leases in the district. Another partnership, Thomas and Manns, owned the Northern Cafe and Bar and the Mint Saloon. Bannock also had a two-story hotel and a small red-light district during its peak in 1909–10. Water for the camp was originally brought from Galena Creek and sold in Bannock for two bits a bucket. Later, the Bannock Water Works was organized and pumped water from Galena Creek to a concrete tank at the Reno Mine. The town folded, however, before operations actually began. Bannock died a quick death. One week the ore was there. The next, the pockets had been emptied. By summer 1910 the camp was almost completely abandoned. The post office closed on July 15, making Bannock's ghostdom permanent. Occasional efforts were made until the 1930s to mine the placers, but nothing substantial was ever found. Today only the placer dumps mark the site.

Battle Mountain (Safford) (Old Battle Mountain)

DIRECTIONS: *Located on U.S. 80, 69 miles west of Elko.*

The town of Battle Mountain came into being in 1867 to serve as a supply center for the Battle Mountain Mining District. The town was originally located just south of Copper Basin. Robert Macbeth named the town in honor of an Indian battle that took place nearby during 1861. The original camp of Battle Mountain was located near the incredibly rich Little Giant Mine, discovered in 1867. When the Central Pacific Railroad came through Lander County (then still part of Humboldt County) in 1868, the camp was moved a couple of miles away, next to the railroad. Battle Mountain quickly became a very important railhead and a well-known dinner station. The Capitol Hotel was the choice eating establishment. The Northwestern Stage Company set up stage lines and ran them to Austin, Geneva, and the Battle Mountain Mining District. By 1870 Battle Mountain had become the supply base for most of Lander County's mining districts.

A post office opened here on June 2, 1870, with Thomas Reagan as postmaster, and the town continued to expand rapidly. Businesses of all sorts opened during the early 1870s. The Martin and Hogan Furnace was started on June 12, 1870. Newspapers also began publication during the 1870s and 1880s. The first was *Measure for Measure,* published by William Forbes. Pub-

At the Battle Mountain depot, with the Nevada Central in the foreground and the Central Pacific in the background. (Nevada State Museum)

lication began on December 26, 1873, and the subscription price was $5 per year. The last issue of the paper appeared on October 9, 1875. The paper folded when Forbes died suddenly on October 13. Another prominent paper was the *Battle Mountain Messenger,* which began publication on May 19, 1877. The paper was organized by Mark Musgrove and sold to E. A. Scott in August. A fire on July 6, 1878, destroyed the building that housed the paper. It was soon rebuilt, and various leaseholders ran the paper until it folded on December 26, 1884. The *Lander Free Press* was a short-lived paper owned by Charles Sproule. The first issue was published on July 1, 1881, but because of a lack of interest in the publication, the *Press* folded on December 29, 1882. Easily the most prominent and professional of these early papers was the *Central Nevadan,* owned by John H. Dennis and first published on January 16, 1885. Dennis leased it to J. D. Park in October 1889. Park became disheartened by the paper's slow progress and gave up his lease in 1890, after which Dennis sold the paper to R. C. Blossom. The *Central Nevadan* eventually merged with the *Battle Mountain Herald,* in December 1907.

By 1880 Battle Mountain had become a well-established town. The population stood at a little over 500, and the Nevada Central Railroad was completed to Austin, bringing additional importance to Battle Mountain. A 300-car siding was built here, along with a large depot, machine shops, and engine house. When Lander County fell into a mining slump during the late 1880s and 1890s, Battle Mountain became the central shipping point for livestock.

The town was the victim of several disasters, including fires and floods. Two fires, in July 1877 and fall 1878, caused $40,000 in damage. In March 1880 another fire destroyed the Huntsman Hotel and the new Nevada Central Railroad depot, causing $15,000 in damage. A very serious flood struck the town in February 1910, destroying the Battle Mountain railroad yards and many miles of the Nevada Central Railroad. Service was not restored until May 30.

Today Battle Mountain is the Lander County seat, having replaced Austin in 1980 after many decades of bitter competition. Battle Mountain is still a center for cattle shipping and mining activity. Because of its location alongside U.S. 80, the town has become a tourist base and one of the main stops along the route between Reno and Salt Lake City. The proximity of the Battle Mountain Gold Company (formerly the Duval Corporation) is a major source of support for the community.

While most of the buildings in Battle Mountain are new, a search of the back streets reveals numerous vintage buildings left from the 1870s. Supplies of all sorts are available here, and the traveler looking for entertainment can visit a number of casinos. A trip to Battle Mountain is very enjoyable, and the historical sites are well worth some attention. If you are lucky enough to be here at the right time, you may experience a unique Nevada treat: During the winter, Battle Mountain has some of the most beautiful "pogonip," a stunning coat of white frost that covers everything in town.

Betty O'Neal (Kimball) (Kimberly)

DIRECTIONS: From Galena Railroad site, take road heading southeast toward Lewis and follow for 3.5 miles. Exit right and follow for 0.5 mile to Betty O'Neal.

Valuable silver deposits at Betty O'Neal were found during the summer of 1881 by prospectors from nearby Lewis. The most important of these initial strikes was the Estelle Nevada Mine, owned by the discoverers, whose names were Blossom, McWilliams, George, Green, Cozzens, and Satler. The following year, in September, the most important mine of the district, the Betty O'Neal, was discovered. The Betty O'Neal Silver Mining Company, with D. P. Pierce as superintendent, was soon organized. Disaster struck the mine on October 31 when the boiler at the hoisting works exploded, killing carman Thomas Bastian and destroying the engine house, blacksmith shop, and hoisting machinery. The workings were immediately rebuilt, but a fire in April 1883 once again destroyed them. That disaster, coupled with the quick decline of nearby Lewis, led to the closing of the Betty O'Neal Mine.

The district remained inactive until late 1907, when the Betty O'Neal was leased by Sherman Wilhelm, manager of the Nevada-Omaha Mining and Milling Company, which also controlled the mines at the short-lived boom camp of Bannock, across the valley. Wilhelm and a small crew narrowly escaped death in the summer of 1908 when a section of mine that they were examining collapsed. Luckily, the men sustained only minor injuries, but Wilhelm never set foot in a mine again for the rest of his life! A small camp formed at the mine, and a post office, named Kimball, opened on April 25, 1910. However, Wilhelm left the district and closed the mine in September 1911. The post office followed suit on October 31, and once again the district was abandoned.

In 1922 one of the giants of the Nevada mining industry, Noble Getchell, purchased the mine from George Abel. Getchell immediately began to expand the mine, adding to its four miles of workings. The Getchell holdings were combined with the Betty O'Neal Mines Company, with George Sias as president. The company controlled eighteen claims (260 acres) and also owned 160 acres of prime ranchland just below the mine. In April the Cahill Brothers of San Francisco were contracted to build a 100-ton flotation mill, which began operations on October 20. In January 1923, Betty O'Neal Mines bought all holdings controlled by the Battle Mountain Mining and Development Company for $1.25 million. The Betty O'Neal company now owned four main veins of ore: Betty O'Neal, Estella, McGarr, and Nebraska. The district reached its peak during the next couple of years, with the most important producers being the Betty O'Neal, Estella, Star Grove, Getchell, and Eagle Sunsag mines.

Although Betty O'Neal was the scene of a lot of mining activity, very few businesses made their home here. Most people chose to live in Battle Mountain because modern transportation made it more economical to be located there. The camp did, however, support a few mercantile stores, a short-lived newspaper (the *Concentrator*), and a baseball team. The team was made up of tough miners with a reputation for playing baseball as a contact sport! However, they were the champions of the Northern Nevada Baseball League a couple of times, so their roughness must have paid off. The post office reopened on June 22, 1925, with the name Betty O'Neal, and Alta Ashcroft was postmistress.

The mines continued to produce, and the mill was expanded to 250 tons to handle the increased ore loads. By 1928, however, ore values began to slip. The low price of silver in 1929 forced the company to cut back to only 12 men. The mine was operated only intermittently until 1932, then was shut down for good. The post office closed on April 26, 1932, and the camp was left to the ghosts. Between 1881 and 1932, ore valued at $2.4 million was produced. Today the mine has been reactivated, and the new company is also reprocessing the expansive ore dumps. Not much remains at the site. The town is off limits to visitors unless they obtain permission from the mine owners. The mill ruins are all that remain from the earlier activity.

Bridges

DIRECTIONS: From Hot Springs, head south on Nevada 305 for 12 miles to Bridges.

Bridges was the site of a six-car siding on the Nevada Central Railroad and was used from 1880 until the railroad folded. The siding was named for Lyman Bridges, president of the Nevada Central. Nothing remains at the site except sections of washed-out railroad bed.

Buckingham Camp

DIRECTIONS: From Copper Basin, continue on for 1.5 miles to Buckingham Camp.

Buckingham Camp was an active mining district during the 1920s and 1930s. Silver ore was first discovered there in late 1918 by Axel Johnson, who organized the Buckingham Mining Company in March 1919. The company controlled 27 claims in the district. Developments included two mines

The Bridges station being dismantled in 1940. (Nevada Historical Society)

that followed a vein of ore varying from 2 feet to 15 feet in width. While production was not extensive, it was consistent. A small camp formed, but most of the company's employees lived in Copper Basin. In May 1925 a new company, Buckingham-Mina Consolidated Mines, entered the district and began extensive mining work. In June 1926 construction on a mill was begun, and in December the 50-ton mill was started, employing 15 men. It was equipped with a 320-horsepower Bolinder engine and an electric generator, supplying 450 volts of power to the mine, mill, and camp.

In June 1929 the mining companies of the district consolidated, and the Buckingham Mines Corporation was formed. The mill was enlarged to 120 tons and a 4,000-foot electric tramway was constructed to bring ore directly from the mines to the mill. The corporation also branched outside of the Buckingham District and took an option on the Hilltop gold mine (Lander County) from the Hilltop-Nevada Mining Company. At Buckingham, the corporation controlled 34 claims, covering 600 acres. The main mine, the Buckingham, was a 1,000-foot incline shaft with many extensive drifts. At the 1,000-foot level, more than 1,000 feet of drifting had been dug. All told, the Buckingham had more than 6,000 feet of lateral workings. The ore supply, however, diminished rapidly after only one year. The mill closed in late 1930, and the corporation waited in vain for a rise in silver prices. That rise didn't come soon enough, and the corporation folded in the mid-1930s. The mines were worked off and on during the following years, with the last activity taking place in the early 1970s. Today the Amax Corporation has been conducting molybdenum exploration in the district. Mining remains, including the substantial remains of the mill, are still left at the camp.

Bunker Hill (Victorine)

DIRECTIONS: From Kingston, continue on for 2 miles to Bunker Hill.

Initial ore discoveries at Bunker Hill were made during the spring of 1863. The three most important mines were the Victorine, the Bi-Metallic, and the Gold Point. The Gold Point, also known as the Iroquois, was located by a Dr. Goodfellow in July 1863. A mining district was organized, but controversy began to cloud ownership of claims in the district. In the spring of 1864, a new mining district, the Bunker Hill, was formed from the Summit and Santa Fe districts by the Smoky Valley Mining and Agricultural Society. The society received the uncomplimentary nickname of "the jumpers" because it relocated claims in the Bunker Hill District that had been challenged by others in the old districts. J. C. Herrin and his mining company gained control of the disputed claims, and the recorder of the Summit District, R. Y. Anderson, filed suit against Herrin, claiming that claims of the Bunker Hill

District were not valid. However, it was discovered that Herrin had acquired all the claims legally. It turned out that Anderson and Goodfellow had caused most of the troubles. They had sold the claims, then turned around and tried to negate the sale by filing suit and keeping the money from the sale. The courts saw through the charade, and Herrin was awarded the decision.

In 1864, Herrin and Company built a 20-stamp mill at Kingston that treated ore from the Gold Point Mine. With a mill nearby, exploration around Bunker Hill increased, and new mines opened, including the Brown, Real Del Norte, Phoenician, Stephens, Empire, Morning Star, Jackson and Green, and Mountain Boy. The growth in activity led to the formation of a small camp. Stone cabins were built, but before substantial development could take place, Kingston became the main settlement. After that, Bunker Hill was mainly just the scene of mining activity and had very few permanent inhabitants. In March 1865, the Starr King Mine joined the list of Bunker Hill's producers. The Sterling Silver Mining Company, with T. L. Burton as superintendent, owned the mine, which produced ore valued at $275 per ton. In August 1867 the Bunker Hill Mill (also known as Coover's) began operations, treating ore from the Victorine Mine. The mines were active until the Big Smoky Mill at Kingston closed in 1869. The mines were operated intermittently afterwards, then completely abandoned in 1887. A few stone remains still exist at Bunker Hill, along with mine and mill ruins. The setting is absolutely fantastic, with a wooded campground and a cool, clear creek. Plan to stay a day and night here, if not for the history, at least to enjoy the beauty of the Bunker Hill area.

Burro

DIRECTIONS: *From Grass Valley, head south on Nevada 306 for 7.5 miles. Exit right and follow for 2 miles to Burro.*

Burro was a small mining camp that sprang up during the spring of 1906. Anthony Dory located the Valley View Mine and served as postmaster when the office opened on July 19. A small camp of 25 formed, but the boom went bust before the end of the year. The later ore shipments never showed the same value as the first few tons, and the small ore pocket quickly vanished. Dory continued to work the mine, but everyone else left the camp. The post office closed on March 16, 1907, and Dory left soon after. The only activity after 1907 took place in 1932, when Gus Laurent worked nine claims near the Valley View Mine. After a few small shipments of quicksilver, Laurent gave up. The district has been abandoned ever since. Not much remains at Burro, only a couple of stone ruins, which are quite difficult to locate. To find the site, look for a small mine dump; the ruins are nearby.

Buzanes Camp

DIRECTIONS: *From Brown's Station, continue west on Nevada 2 for 6 miles. Exit left onto poor road and follow for 3 miles. Exit right and follow for 1 mile to Buzanes Camp.*

John Buzanes, who was from Greece and had been working in mines at Cripple Creek, Colorado, discovered gold in 1926. The ore, gold in iron oxide, assayed at an average of $17.20 per ton. In September 1927, Buzanes sold his property to the newly formed Magna Gold Mines Company for $200,000. The company, which owned 34 claims covering 1,342 acres, was incorporated in Magna, Utah, with Peter Athos as president. The main development was a 200-foot mine, with 2,000 feet of lateral work. A couple of smaller shafts were also dug but were never very productive. A small camp of 15 formed and five buildings were constructed. A mining office, an assay office, two bunkhouses, and a cookhouse formed the core of the camp. The company built a small Straub-type mill that began operations in April 1928. However, ore values fell dramatically during 1929, and it was no longer profitable for the company to continue mining. Magna folded in 1930, and no other activity has taken place at Buzanes Camp since that time. Today one of the bunkhouses still stands, but the only other remains are a couple of collapsed wooden buildings. The road to the site is quite rough, so exercise caution.

Canyon (Reese River Canyon)

DIRECTIONS: *From Walters continue north on Nevada 376 for 3 miles to Canyon.*

Canyon was the site of a sixteen-car siding on the Nevada Central Railroad. The siding was active from 1880 until the closure of the railroad but never achieved any prominence, and nothing at all remains today.

Canyon City (Big Creek) (Lander City) (Watertown) (Mineral City) (Montrose) (Middletown)

DIRECTIONS: *From Austin, take U.S. 50 west for 2.5 miles. Then take Nevada 2 south for 6.4 miles. Exit left onto Nevada 306 and follow for 4.5 miles. Exit left, sharp, and follow for 3 miles. Exit right and follow for 5 miles to Big Creek Campground (Canyon City).*

Prospectors from Austin located several claims along Big Creek during the early months of 1863. Small villages were established and later consolidated into one town, Canyon City. The settlement expanded rapidly.

A telegraph line was constructed from Austin to Lander City, located at the mouth of Big Creek Canyon. A post office opened on August 19, with William Logan as postmaster, and Canyon City continued to boom. The *Mining and Scientific Press* reported that Canyon City "would appear to be the most promising mining settlement in the Central Nevada region." By the end of the year, Canyon City boasted a hotel, two restaurants, three saloons, a butcher shop, a notary public, and a recorder's office. Only a year later, the town's population was nearing 1,600. Artemus Ward, famous lecturer and humorist, visited the town in late 1864 and spoke in the Young America Saloon, which wasn't exactly a choice spot. The saloon had a dirt floor and a sagebrush roof, but, despite these conditions, Ward's speech was a huge success. Two large mills were also in operation by 1864: the Eureka (10 stamps) and the Parrott (16 stamps, later enlarged to 20), also known as the Pioneer. There were numerous mines in the district, but by far the most valuable was the Bray-Beulah. The most prominent mining companies were the Gold Point Gold and Silver Mining Company, Monarch Gold and Silver Mining Company, Phuebus Gold and Silver Mining Company, Olympic Gold and Silver Mining Company, Seely Gold and Silver Mining Company, Eureka Mill and Mining Company, and Astor Ledge and Company. The ore in the district was laced with copper, which made refining difficult. Furnaces were constructed, but none ever operated successfully. By the end of 1864, four additional crushing mills, including the 10-stamp Phelps and the 5-stamp Lippett, had been completed and were in full production.

By 1865, Big Creek Canyon was crowded with stone cabins, and residents supported almost twenty stores, a school, a justice court, and express and telegraph offices. The Eureka Mill was improved in November and began processing ore from the Whitlatch Mine, including one shipment valued at $1,800. However, the mines all started to go dry in early 1866, and Canyon City began a quick slide to the bottom. The Eureka Mill was dismantled and moved in June to the Philadelphia District (Nye County). By the beginning of 1867, all mines had closed. Only the Parrott Mill was still in operation, processing ore from Austin, but in September it closed too, signaling the end of mining activity in Big Creek Canyon. The town quickly emptied and by the time the post office closed on October 14, the area was virtually abandoned. Production from the Big Creek Mines from 1863 to 1867 totaled just over $500,000.

During the 1880s, only ranches operated in the canyon. Sam Markwell owned the Hogan Ranch and grew alfalfa (which sold for 1.25 cents per pound), apples, and other crops. The population of the entire district had shrunk to 26 by 1881. It wasn't until 1890, when the Pine Mine was discovered, that mining activity returned to the canyon. In 1891 Joseph Bray relocated the Bray-Beulah Mine and renamed it the Bray Antimony Mine. The ore from Bray's Mine, for no reason this author could find, was shipped to

The town of Big Creek in 1911, long after the boom had ended. A small antimony mine is at the right. (Nevada Historical Society)

Swansea, Wales, for a short while. In July 1892, Simon Bray, Jefferson Hull, James Williams, Joseph Bray, William Falvey, and James McLaughlin consolidated their holdings and formed the Big Creek Mining Company. By October four mines were in operation. One of these was the Androval Mine, located on the side of the mountain. The ore was brought down from the mine via a 1,300-foot rope tramway. Production was never large and only scant monetary gains were realized.

The mines operated until 1898, when the district was once again abandoned. In 1907, the Antimony King Mine was discovered, and it operated off and on until 1922. The Bray Mine was reopened in 1916 and shipped small amounts of ore but closed in 1918. Once the Bray Mine closed, only the Antimony King Mine was active. The mine was bought by the Nichols-Laying Chemical Company of San Francisco in 1922, then sold to J. G. Phelps Stokes of Austin. In 1936 the Big Creek Mining and Milling Company acquired the property, which consisted of four mines: the Mammoth, Mountain View, Confidence, and Commodore. The company also owned the Hard Luck–Pradier Mine, which produced 400 tons of antimony ore before it closed in 1958. The Antimony King was much more valuable, producing 1,120 tons of ore in 1957 alone. Unfortunately, that output emptied the ore pocket, and all mines closed for good in 1958.

Ruins of Big Creek's settlements are scattered throughout the canyon. Stone and brick remains mark the site of Canyon City and Lander City. A new mine has recently been started, so beware of ore trucks. A trip to Big Creek is well worth the time.

Catons

DIRECTIONS: From Silver Creek, head south, following old railroad bed, for 5.5 miles to Catons.

Catons was named for the Caton family, who ran a small ranch nearby. A twenty-car siding, part of the Nevada Central Railroad, was also located here, but no development ever took place. Today nothing remains at the site except some of the old ranch buildings.

Clifton

DIRECTIONS: Located 0.75 miles west of Austin.

Clifton came into being after William Talcott made his ore discoveries in Pony Canyon on March 2, 1862. The Clifton townsite was platted by two men named Marshall and Cole, just below the original district claims. A fair-sized town quickly formed as the rush to the new mining district developed. By spring 1863, a tent city of 500 had sprung up at the mouth of the canyon. Flora Bender, while traveling through Clifton, reported, "The houses are principally of canvas, with roofs made of cedar brushes but very few wooden buildings." The businesses in town included a Wells-Fargo office, assay office, recorder's office, justice of the peace, hotels, restaurants, and lumber and hay yards. As Clifton was becoming a prosperous new town, however, the Austin townsite was platted farther up Pony Canyon.

Austin and Clifton enjoyed a spirited competition. Clifton had a level townsite, but Austin offered free lots to businesses in exchange for help in building a graded road from Reese River Valley to Austin. While Clifton was slightly larger, when the road (which bypassed Clifton) to Austin was completed, the battle was lost. The combination of Clifton's high-priced lots and Austin's free lots spelled doom for the town. When Austin won the battle for the county seat in the fall of 1863 Clifton's fate was sealed. People began to move to Austin and finally, on February 20, 1864, the Clifton post office closed. Soon Clifton was abandoned, with empty wooden buildings scattered over the site, little islands in what had once been a sea of tents. A small 4-stamp mill, the Clifton, was active in 1865, but interest in Clifton was gone, and when the mill closed in 1867 Clifton was left to the ghosts. Nevertheless, a post office, with Frederick May as postmaster, ran from June 20, 1867, to September 28, 1868. But not even the postal system showed much interest in Clifton.

Clifton was revived in 1880 when the town, now known as Austin Junction, became the terminus for the Nevada Central and Austin City railroads and a thirty-car siding for the Nevada Central was constructed. The railroad connection was the sole purpose of the town, however, and no people moved

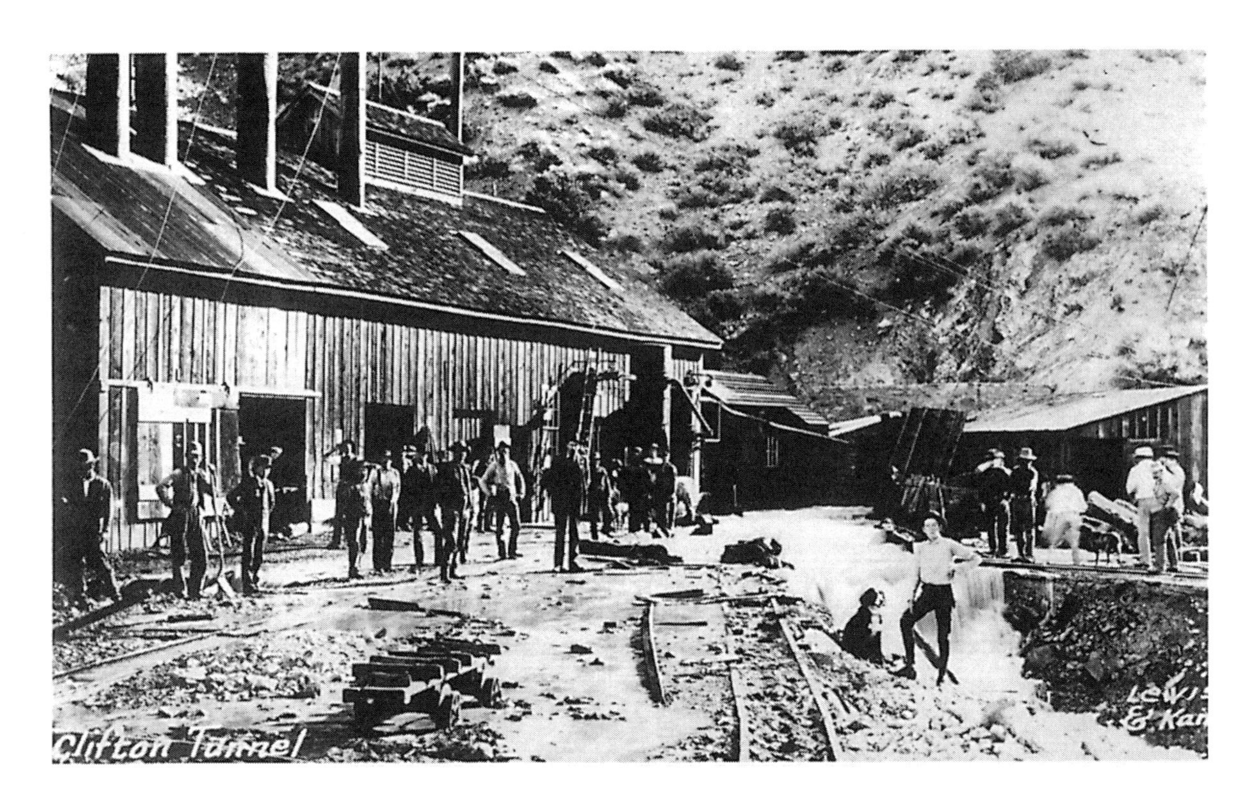

Mouth of the Clifton tunnel shortly after its completion.

back as residents. In 1894, J. G. Phelps Stokes, president of the Austin Silver Mining Company, constructed the 40-stamp Clifton Mill at the mouth of the 6,000-foot Austin Tunnel. The mill, equipped with machinery from the Manhattan Mill in Austin, operated for more than a decade. However, Clifton never had any permanent residents during this period. Austinites used the area for recreation; a baseball diamond was built, and townspeople enjoyed weekend picnics at the old Clifton townsite. Today remains are scant. The mill ruins are about the only markers of the site.

Clinton

DIRECTIONS: *From Austin, take U.S. 50 east for 11.5 miles. Exit right and follow Nevada 376 south for 3 miles. Exit right and follow rough dirt road for 2.5 miles to Clinton.*

A small group of prospectors located some promising silver deposits during the summer of 1863, and a townsite was soon platted. In January 1864 a letter sent to the *Reese River Reveille* said that Clinton will be "one of the most prominent towns around here." For a while, prospects for Clinton looked very good. A post office, with George Ellery as postmaster, opened on

April 23. Numerous mines began operations, the most prominent of which were the Antiquarian, Chieftain, Comstock, Ellery, Clinton, Gold Eagle, Great Central, and Masonic. The Antiquarian was the most valuable, yielding ore in June that assayed at $1,000 per ton. Ellery was heavily involved in the district's mining affairs. He served as the superintendent of three mining companies: Antiquarian, Comstock, and Ellery. Clinton, however, faded quickly. The ore deposits played out, and when the post office closed on December 23, 1864, the small camp began to empty. A small 4-stamp mill was built in Clinton during the spring of 1865. The mill was a failure and operated only periodically before it closed for good in 1866. No activity has taken place since. Stone ruins mark the site today.

Copper Basin

DIRECTIONS: From Battle Mountain, head south on Nevada 376 for 5 miles. Exit right and follow for 3.5 miles to Copper Basin.

Copper Basin came into being in 1897 when the Glasgow and Western Mining Company began operations in the district. A small camp of about 50 formed. The two main mines, the Turquoise King and the Myron Clark, were the early best producers. Glasgow and Western continued to operate off and on for about a decade before closing down. The claims were taken over by the Copper Basin Mining and Milling Company, with L. G. Hardy as president. The company then controlled six claims, developments that included 1,200 feet of lateral work on a 250-foot shaft and a 300-foot tunnel. In 1917, the Copper Canyon Mining Company purchased all Copper Basin mines—120 claims (2,172 acres), 20 placer claims (402 acres), 17 mill sites, and a 280-acre townsite.

The company developed 5 of the mines: Widow (200 feet, 2,900 feet of lateral work), Chase and Goff (200 feet, 3,045 feet), Sweet Marie (295 feet, 4,000 feet), Contention (130 feet, 2,700 feet), and Hawkeye (403 feet). Copper Basin's peak came between 1918 and 1920. The mines were idle in 1920, but $76,000 of exploration work was conducted in 1924. Leaseholders then worked the claims until 1935, when the Copper Canyon Mining Company regained control of the mines. The company did no mining work, though, and in 1941 the property was leased by the International Silver and Refining Company. The Copper Canyon Mining Company once again regained control of the mines after the war. The mines were worked intermittently until 1957, when the company left the district for good. Then the district was idle until 1967, when the Duval Corporation, now known as the Battle Mountain Gold Company, reopened the mines. The company is very active in Copper Basin today. A couple of wooden shacks from the old site remain, but most of the original buildings have succumbed to the company's recent expansion.

Copper Canyon (Copper) (Natomas)

DIRECTIONS: From Battle Mountain, head south on Nevada 376 for 12 miles. Exit right and follow for 3 miles. At fork, bear right and follow for 3 miles to Copper Canyon.

Initial discoveries in the Copper Canyon district took place during 1871. An English company gained control of the main mines and shipped some 40,000 tons of ore to Swansea, Wales. Although the ore was valuable, shipping costs drove the company out of the district in the late 1870s. It wasn't until the Glasgow and Western Mining Company bought all the copper mines in 1897 that Copper Canyon began to reach its peak. Of the several mines brought into production, the Virgin and Lake Superior mines were the most important. A camp of 50 grew at Copper Canyon. On May 3, 1906, a post office named Copper, with Donald McDonald as postmaster, opened at the camp. While the post office closed later the same year, on November 15, the camp continued to prosper. In 1909 substantial placer deposits were located at the mouth of Copper Canyon and a new camp, called Natomas, sprang up there.

Mining operations at Copper Canyon, with the town to the left.

The Natomas dredge, the small camp's lifeline, in full operation. (Nevada Historical Society)

In 1916, the Copper Canyon Mining Company acquired most of the claims in the Copper Canyon and nearby Copper Basin districts. The company, which was incorporated on November 4 in Delaware, controlled the Copper Canyon Group, 19 claims covering 160 acres. The three major veins were the Virgin, the Superior, and the Eckes. While Copper Canyon Mining was the largest company in the district, several others were also active during the 1910s and 1920s. The Toulon Arsenic Company owned the Irish Rose Mine and shipped its ore to a 150-ton concentrator in Toulon (Pershing County). The company was active for a few years but left the district during the mid-1920s. The Iron Canyon Gold Mining Company worked the Buzzard Mine and produced $80,000 before leaving. The Dahl Placers, located at Natomas, were also important producers. James Dahl controlled 25 claims, and from 1910 to 1919 $485,000 in gold was removed.

By the end of the 1920s, the Copper Canyon Mining Company was the only active company. From 1916 to 1940 the company produced over $1.6 million. The property was leased to the International Silver and Refining Company in 1940. The company built a 350-ton mill, but it was never a success. The mill closed in 1944 and the company left the district soon after. During the 1940s, three other companies (Barium Products Limited, London Extension Manufacturing Company, and California and Nevada Barite Company) worked the district without much success. The Natomas Gold Dredging Company was organized in August 1949. The company brought in a huge dredge that had been working in Manhattan (Nye County) and had favorable returns for a couple of years. The Copper Canyon Mining Company became the sole operator in the district until 1957 when it too folded. The mines were abandoned until 1967, when the Duval Corporation began operations. Now known as the Battle Mountain Gold Company, the company is still very active in the district and has had impressive gold production. Today, mill ruins are all that's left of the original settlement. Virtually all of the old wooden buildings were removed during recent mining expansion.

Cortez

DIRECTIONS: *From Tenabo, head directly east for 4 miles, crossing paved road at 0.5 mile. Exit right and follow graded road for 12.5 miles. Exit left and follow for 1 mile to Cortez.*

The long history of Cortez begins in 1862 when a group of Mexicans discovered silver ore next to Mount Tenabo. The ore was shipped to Austin, where it raised considerable interest among the locals. An eight-man prospecting party, led by Simeon Wenban, traveled to the district in March 1863 and found silver float in the gulches southwest of Mount Tenabo. Wenban and his group followed these floats and located substantial silver ledges, which sustained Cortez for many years even though the isolated location of the deposits made development difficult. In 1864 Wenban formed a partnership with George Hearst, father of William Randolph Hearst, and that summer the first mill in the district was built. An 8-stamp mill, the Cortez, was constructed in nearby Mill Canyon, but its success was limited and most ore was still shipped to Austin.

The troubles with processing and the camp's site did not dampen Cortez's growth, however, and by 1865, three mining companies were active in the district: the Cortez Company (the largest, Wenban's company), the Edward Everett Silver Mining Company (C. H. Burton, secretary), and the Pizarro Silver Mining Company (also with C. H. Burton as secretary). The Cortez company spent $100,000 in 1865 to enlarge the Mill Canyon Mill to 16 stamps in response to the huge volumes of ore being produced from the Garrison Mine, the richest of the early mines. The camp's population swelled to more than 400, and close to 200 men were employed just to cut the wood needed to keep the mill running around the clock. A post office, with James Russell as postmaster, opened on January 3, 1868. Hearst, who was not happy with the progress of the mines, sold out to Wenban and left the district for good.

The town of Cortez in 1900. The Cortez Smelter is in the foreground. (Nevada Historical Society)

The discovery of the St. Louis Mine in 1868 gave Cortez an additional boost. An 88-ton test shipment from the mine had values of $600 per ton. While the post office closed on October 15, 1869, activity continued to expand, and Wenban gained control of virtually all the claims in the district. For the next two decades, the Aurora, Arctic, Benjamin Harrison, and Garrison mines proved to be the district's mainstays.

In 1886, Wenban built a 50-ton leaching plant at the Garrison Mine. The mill continued to operate until 1894. From 1887 to 1891, the Cortez mines had an impressive record, each year producing over $300,000. The post office reopened on June 25, 1892, and continued to operate until June 15, 1915. Cortez's founder, Simeon Wenban, died in 1895, and from 1895 to 1919, only leaseholders worked the district. Production, however, was still impressive. In 1919, big changes came to Cortez. The newly organized Consolidated Cortez Silver Mines Company, with F. M. Manson as president, gained control of the district's mines. The company purchased 49 claims consisting of the Cortez group and the water rights and mill sites formerly owned by Tenabo Milling and Mining Company. Consolidated Cortez employed 150 men, and in 1923 built a 150-ton cyanide mill.

Because of the large work force, the post office was reopened on January 3, 1923. Mining activity continued to increase during the 1920s, with the Garrison Mine continuing to be the main producer of the district. By 1924 the Garrison Tunnel was 4,000 feet long, connected on the fifth level to the St.

Part of the mine crew at Cortez. (Nevada Historical Society)

Louis Mine and on the sixth level to the Fitzgerald Mine. The other big producer was the 2,170-foot Arctic Tunnel. By the late 1920s and early 1930s, however, silver prices dropped dramatically and activity in Cortez slowed down. The mill was shut down in January 1930, and the company finally folded a couple of years later. By the time the mines were closed, the Garrison was 4,500 feet long and 1,270 feet deep, with ten levels and more than fifteen miles of workings. Production from 1919 to 1929 was just over $2 million.

After Consolidated Cortez folded, only leaseholders were active in the district. In 1959 the American Exploration and Mining Company took over Cortez's mines, which had a total production of more than $14 million from 1862 to 1958. In 1969, the company built a 1,500-ton cyanide mill that produced $6.6 million in its first year of operation. Today the property is controlled by Placer Amex Incorporated, and the mining and milling operations employ 150 workers. Discoveries recently made in nearby Horse Canyon revealed deposits estimated at 3.4 million short tons with a projected annual gold production of 40,000 ounces. Placer Amex has taken a refreshingly sincere interest in the history of the area, a much different attitude than that of companies mining in other historical districts. For example, when expansion dictated that the main roads to the mill (which go through what used to be downtown Cortez) be widened, the company was careful to make sure that none of Cortez's remains were disturbed. We can only hope that such an example of history and mining working together will be taken to heart by other mining companies. Shaky wooden buildings and scattered ruins remain at the Cortez townsite, located a couple of miles east of the new Cortez Mill.

Curtis

DIRECTIONS: From Silver Creek, continue north for 3 miles to Curtis.

Curtis was one of the largest sidings on the Nevada Central Railroad, with forty cars. The siding was named for A. A. Curtis, treasurer of the Nevada Central. Nothing, aside from some railroad-related buildings, was ever constructed. Today only the wooden remains of a small collapsed building mark the site.

Dillon

DIRECTIONS: From Galena Railroad, continue following the Jenkins Road for 4 miles to Dillon.

Dillon was a small railroad siding located at milepost 14 of the Nevada Central Railroad and named for Sidney Dillon, a railroad executive. During 1916, small copper mines in this area sent ore through Dillon. After closing in 1917, Dillon faded into obscurity. Today nothing at all remains at the site.

Dry Creek Station (Capehorn Overland Stage Station)

DIRECTIONS: From Austin, take U.S. 50 east for 22 miles. Exit left and follow good road for 2.75 miles. Exit left and follow for 3 miles to Dry Creek.

Dry Creek was a stop on both the Pony Express and the Overland Stage routes. The station was built in the spring of 1860, making it one of the last stations built by the Bolivar Roberts division of the Pony Express. Dry Creek was the scene of constant Indian troubles, the most horrible of which occurred on May 21, 1860. Sir Richard Burton reported the following:

> At the time of the fight, there were four men at the station: Silas McCanless, the station-keeper, John Applegate, Ralph M. Lozier, and W. L. Ball, the pony express rider. McCanless was living with a squaw and it appears that the Indians were dissatisfied with this fact, and wanted the squaw to return to the tribe. Early in the morning of the fight, the Indians, numbering about fifteen or twenty, who were camped near by, came to the station and demanded of McCanless to give up the squaw. Considerable wrangling and high talk was engaged in, but she was not given up, and McCanless, having given the Indians a generous supply of rations and in a manner pacified them, they went off evidently satisfied. They returned, however, at about seven O'clock, and creeping

> up to the station, which was built of cottonwood logs, and being newly constructed, had not been "chinked" with logs, and at the first volley, killed Lozier and severely wounded Applegate, he being shot through the fleshy part of the thigh, the ball ranging up and coming out through the pocket in his pants. Leaving Lozier dead in the station, the three men, McCanless, Applegate and Ball, fled from the place for dear life, with the Indians in hot pursuit. Applegate, at the outstart, had handed his revolver to Ball. After running about a quarter of a mile, McCanless' squaw in the meanwhile running between them and the Indians, and endeavoring to keep the latter back, Applegate, who was badly wounded and was fast failing from loss of blood, knew that he could not hold out in the race, and halting he asked Ball for the revolver, and rather than be overtaken by the Indians, who were close upon them, and dreading the torture they would inflict, placed the pistol to his ear and deliberately blew his brains out. McCanless and Ball continued to run for their lives. In order to lighten themselves they fairly stripped to their underclothing, and after a most desperate flight of several miles managed to outstrip the Indians, who gave up the chase. The two men continued on until they reached the station at Robert's Creek, thirty miles distant from Dry Creek.

After that incident, a new stationmaster, named Colonel Totten, was appointed. He remained until the station closed. The completion of the telegraph in 1861 reduced the need for the Pony Express. After the Pony Express folded, Dry Creek served as an Overland Stage stop from 1861 to 1869. After that station closed, Dry Creek reverted to being a ranching operation, which is still active today. There are very few remains of the two stations. A few overgrown stone foundations near the ranch mark the Pony Express station, while a partially standing stone structure is all that remains of the Capehorn Overland Stage station, located six miles to the southwest. A commemorative plate was placed here in 1960 by the Pony Express Centennial Commission. The Damele family lives at Dry Creek and has owned the ranch for almost 100 years.

Frisbie

DIRECTIONS: *From Cortez, backtrack 1 mile to main road. Exit right and follow for 3.5 miles. Exit right again and follow for 2 miles to Frisbie.*

Frisbie was a small mining camp active during the 1880s. Initial discoveries were made by miners from nearby Cortez during spring 1883. A camp of 25 formed, and on July 16 a post office, with Charles Harvey as postmaster, opened. Frisbie became the principal mining camp of the Campbell (Bullion) Mining District. Jim Campbell, founder of the district, owned the Limestone Mine in Frisbie. The Twis brothers owned two other good producers, the Osceola and the Bismark mines. Other mines in the district included the Silverside, Lady of the Lake, and Lady Francis mines. While the post office closed on October 29, 1885, the district continued to be active, albeit at a much slower pace. Charles Engstrom, a prospector from Austin, shipped silver ore to Austin's mills in August 1891, producing more than 150 ounces of silver. A school was built in 1891 to serve the families of the miners in Frisbie and Cortez for the next two years. The camp faded during the early 1890s when ore values from the mines dropped, and nearby Cortez began to boom again. By the turn of the century the district was abandoned, and no activity has taken place there since. All that remains of Frisbie is a couple of stone ruins.

Galena (Blanco)

DIRECTIONS: *From Battle Mountain, head south on Nevada 305 for 9.4 miles. Exit right and follow for 3 miles to Galena.*

While initial discoveries in Galena Canyon were made in 1863, no camp formed until 1866. The principal early mines in the district were the Avalanche, Buena Vista, Butte, Cumberland, Evening Star, Ida Henrietta, Lady Carrie, Trinity, and White. By 1868, more than 100 residents lived in Galena. A townsite was platted in 1869, and soon the streets of the town were crowded with mercantile stores, saloons, and other business establishments. A town hall and town water system were also constructed. In 1870, the boom in Galena began in earnest. During that year, two smelters, of 12 and 20 tons, were built. The Nevada Butte Mining Company, owners of the Butte Mine, built a 20-stamp mill, which started on June 19, 1871. A post office, with B. F. Blossom as postmaster (a position he held for twelve years), opened on June 2, 1871. By October 1873, Galena had a population of 250, including 100 miners, supporting two hotels and four mercantile stores. Two stage lines from Galena to Battle Mountain were established, the Tuller and Cluggade Stage ($3 each way) and the Flippini Stage.

Mining activity in the district continued to increase during the early and mid-1870s. The White and Shiloh Consolidated Silver Mining Company, owner of the Battle Mountain, Shiloh, and White mines, built a $60,000, 50-ton concentrating mill in 1875. The company produced $450,000 before folding in the early 1880s, when mining activity slowed dramatically. Despite the reduced activity, the 1881 census showed 348 residents in the town. In 1886 a French company, the Blanco Mining Company, purchased most of the mining property in the district. As a result, the post office was renamed Blanco from 1887 to 1888. However, Galena was restored as the name on October 11, 1888, after a long and loud protest by the citizenry. In September 1889 the mill and assay office of the Blanco Mining Company burned, causing $25,000 in damage and effectively putting an end to Galena's mining activities until World War I. The post office was kept open until November 15, 1907, when it was decided that Galena had faded too much to warrant a post office.

Three mining companies were active in the district during the 1910s and 1920s. The Nicklas Mining Company (E. W. Breitung, president) entered the district in September 1916. The company's main holdings were the Nicklas and Plumas mines. A 30-ton concentrator was built in 1919 and concentrates sent to Midvale, Utah. The Joyce Mining Company (J. C. Hensen, president) began operations in 1917 and worked four claims. The company was not a success and folded in December 1920. In 1919 the Nevada Silverfields Mining Company was organized and acquired the holdings of the Silverfields Mining Company and Nicklas Mining Company. The company was active in the Galena District until 1927, when it folded. Since that time, there have been two revivals. One occurred just before World War II and another in the late 1960s and early 1970s. Total production figures for the district are just under $6 million. Today Galena is a quiet settlement with a population of about 10. Remains from its heyday are scattered throughout Galena Canyon. While only a few wooden buildings from the period still stand, the mining ruins are much more extensive. Mines dot the landscape, and mill and smelter ruins still exist. Galena, and the canyon in which it lies, are a definite must to visit. While the history and sights of Galena are well worth the trip, the beauty of Galena Canyon adds a special extra.

Galena (Lewis Junction)

DIRECTIONS: *From Battle Mountain, head east for 0.1 mile and exit south onto the old Jenkins Road. Follow for 9.5 miles to Galena, located .125 mile west of road.*

Galena was a twelve-car siding on the Nevada Central Railroad. The stop was located on the present-day Marvel Ranch, where a small water stop and station were built. The stop was active as a shipping point during the boom at nearby Lewis but was abandoned when the railroad folded in 1938. Today a small building remains at the ranch, a relic from Galena's railroad days.

Geneva

DIRECTIONS: *Located 0.5 mile east of Clinton.*

Initial discoveries at Geneva were made during March 1863. A seven-man prospecting party, led by Charles Breyfogle, located rich silver and gold deposits on the east slope of the Toiyabe Mountains. Shortly, in November, the Geneva townsite was platted. A rush to the area developed, and by 1864, Geneva had a population of 500. Businesses, including a store, a saloon, a blacksmith shop, and a stage office, opened. Within months, stores and log cabins crowded the canyon. Breyfogle built a hotel, but he was too susceptible to wanderlust and soon left to continue prospecting. Shortly after leaving Geneva, he discovered the famed Lost Breyfogle Mine in Death Valley and ended his days trying to find the mine again. The mines in Geneva continued to boom. In 1865, the Big Smoky Mining Company built a 20-stamp mill, after a rich ore strike on the Smoky Valley Ledge in April. The ore from the Big Smoky Mine originally assayed at $300 per ton. The mill, however, was a waste of money and operated for an extremely short period before low ore values forced its closure.

Despite this setback, Geneva continued to prosper. Other mines in the area, among them the Everett (owned by the Everett Mining Company), Illinois, and Mammoth mines (Smoky Valley Mining Company), also helped to keep ore production high. A post office opened on June 20, 1867, and G. B. Moore was appointed postmaster. But Geneva's hopes for permanence were crushed when the mines gave out during late 1867. The post office closed on September 20, 1868, and the camp soon was empty. All of the machinery from the mill and mines was removed, leaving stone cabins as the only site markers. A small revival took place beginning in 1916, when John Cahill discovered the Smoky Valley Mine near the old Big Smoky Mine. He sold the mine to a newly organized company, the Nevada Birch Creek Mining Company, in 1919. The company was based in Austin, had J. F. Bowler as president, and worked eight claims, including two tunnel mines. Nevada Birch Creek folded in 1921, and Geneva has been a ghost town ever since. Total production for Geneva was a little over $100,000. Today interesting stone cabins, in various stages of ruin, still mark the site.

Gold Acres

DIRECTIONS: *From Tenabo, backtrack 1.5 miles to paved road. Exit right and follow for 5 miles to Gold Acres.*

Gold Acres was one of the more recent mining camps to become prominent in Lander County. The Gold Acres Mine was discovered in 1936, and a mill was built nearby. The property was controlled by the Consolidated Mining Company. An additional 400-ton cyanide mill was constructed in 1941. Between 1936 and 1940, the mine produced $213,000. In March 1942 the property was purchased by the London Extension Company, which employed 100 men. The population of Gold Acres was around 300 during the 1940s, and a school opened in 1942. Close to forty houses were built and businesses opened, including two mercantile stores and a saloon. In October

The scant ruins at Gold Acres.

1945 a Marcy ball mill was moved here from Colorado, and 35 more men were employed. This 400-ton mill did not enjoy the success of Gold Acres' other mill, and it closed down after only one year. While the Gold Acres Mine was the main producer, a number of smaller mines, including the Turquoise 50 ($5,000), Blue Fern ($5,000), and Steirich ($15,000), also produced ore, mainly turquoise. The Gold Acres variety was of extremely high quality.

The Gold Acres Mine was a very consistent producer, and the town was a successful and happy company town. Operations continued until June 1961, when the company folded and Gold Acres was abandoned. The townsite was sold at a sheriff's auction during the fall and in December, all buildings in Gold Acres were bulldozed. Only one person still resides at Gold Acres: Orvil Jack, a well-traveled mine mechanic in his seventies. His machine shop is in an old boxcar that rests on a hill overlooking the remains of Gold Acres. Orvil owns several claims in Gold Acres. His expertise is in mine equipment repair, and he has spent most of his time repairing equipment in Crescent Valley, Cortez, and Tenabo. He believes that his claims have some of Nevada's richest deposits of gold, silver, and turquoise. From the evidence of recent mining reports, he could be right. The Gold Acres district has already produced a little over $10 million. With new activity in nearby Tenabo and Cortez, there is a good chance that the Gold Acres mines might also be reopened. Today only a huge "glory hole," foundations, and wooden rubble mark the site.

Grass Valley (Spencer)

DIRECTIONS: From Austin, take U.S. 50 east for 5.7 miles. Exit left onto Nevada 306 and follow for 19 miles to Grass Valley Ranch.

Grass Valley has had a rich livestock history since 1873. John Spencer, who made his fortune in the Nevada mines, founded the Grass Valley Ranch and lived here for twenty years. Spencer also owned the Birch Creek Ranch in Smoky Valley. He built a huge house and made Grass Valley a profitable ranching operation. When Spencer died in 1893, a long feud developed between his two sons over ranch ownership. A post office opened here on February 20, 1876, and served many ranches in the valley until it closed in 1913. Sarah Spencer, widow of John, served as postmistress. A stage line, the Austin–Grass Valley Stage, was set up during the 1890s. The stage was driven by John Callaghan, whose father, Dan, originally settled the nearby Callaghan Ranch in 1864. The feud between the Spencer brothers was never resolved, and in 1904, the ranch was bought by Walter Magee at a public auction. Magee sold the ranch in 1908 to the Lander County Livestock Company, run by George Wingfield, the Nevada mining magnate. The ranch served as a freight station for ore being shipped from Cortez to Austin. In 1917, Wingfield sold the ranch to John Saval, a Basque sheepman. Saval was killed in an auto accident in 1931, and the ranch passed to George Watts. In 1934, the ranch was purchased by Richard Magee. His former wife, Molly Knudtsen, is the current owner. The huge house that Spencer built burned down in 1951. Several older buildings remain at Grass Valley. Molly Knudtsen is one of the great women of Nevada. Besides being a history buff and an author, she has been very active in many public service organizations, including the Nevada State Museum. Her books provide a fascinating in-depth look at some of Nevada's ranching history.

Guadalajara (Sante Fe)

DIRECTIONS: From Austin, head east on U.S. 50 for 11.5 miles. Exit right onto Nevada 376 and follow for 12 miles. Exit right and follow for 1 mile to the Schmidtlein Ranch. Ask for permission here to visit Guadalajara, located 1 mile above the ranch.

The Sante Fe District was organized by Peter Brandow, Robert Stuart, and John Reed in March 1863 after Mexican prospectors had located substantial silver ledges here. By summer, 3,000 mining claims had been staked in the district. A city site was platted in early 1864, and the boom was on. Among the more notable of the many mines in the district were the King (the most valuable), Mother Lode (owned by the Centary Mining Company, which built a mill in nearby Kingston Canyon), Santa Maria (the original

A Guadalajara ruin showing rifle ports.

location of the district), Otho, Mammoth, San Francisco, Florida, Maryland, Hudson, Amazon, Eureka, and Rattler.

While close to twenty mining companies were active in the area, only a few actually turned a profit. The best-producing mining companies of the district included the Maryland Silver Mining Company, Imperatrice Silver Mining Company, Belle Silver Mining Company, Santa Maria Silver Mining Company, Vulcan Mining Company (Alpheus Bull, president), United States Gold and Silver Mining Company (T. L. Bibbins, secretary), and Taurus Mining Company (C. H. Burton, secretary and recorder for the Santa Fe District).

By summer 1864, Guadalajara had reached its peak. About 150 people lived in the town, and the usual assortment of businesses thrived there. However, long life was not in the cards for this boom camp. In most of the mines, the ore was there one day, gone the next. One by one, the mines closed, and the population began to shrink. By the late 1860s, only a few mines were still producing, and only small amounts of ore were emerging. By 1870, only 50 residents were left. Two years later, only two mines, Yosemite and King, were being worked. A handful of residents still lived here, however, most employed in the mines and mills of surrounding districts. The population still stood at 37 in 1881, but virtually all mining activity within the district had ceased. By 1890 Guadalajara had joined the ghosts. Today the site is one of the most intriguing in Lander County. Close to twenty stone ruins remain, some in surprisingly good condition. Many of the buildings exhibit unique design. A number have narrow slits in the walls for rifles, just in case of Indian attack. This is a fascinating ghost town, but since it *is* on private property, please ask permission at the ranch before venturing up Sante Fe Canyon.

Gweenah

DIRECTIONS: From Ledlie, continue north for 1 mile and exit left. Follow this poor road for 2 miles to Gweenah.

Gweenah came into being in early 1908 when James Watt of Austin discovered the Watt Mine in February, and a small tent camp sprang up next to the nearby camp of Skookum. The Gweenah boom was short-lived, but during 1909 the two camps had a combined population of 200. While Gweenah folded the next year, the Gweenah Mine, located by the Lemaire brothers in 1909, continued to produce until the late teens. Total production for the Gweenah mines was just over $15,000. The town did not exist long enough for any permanent structure to be constructed, and so today only mine dumps mark the site.

Hilltop (Kimball) (Marble City) (Marble Canyon)

DIRECTIONS: From Tenabo, head north for 1.5 miles. At the end of road, exit left and follow for 12.5 miles. Exit right and follow for 2 miles to Hilltop.

Long before mining came to this district, the canyon area was active. A camp named Marble City was located here during the 1860s and served as a stop for travelers, a watering place for horses, and a place to get a good meal. Some mining exploration was undertaken, but no major work was done until 1887, when the Mayesville Mining Company worked the Mayesville Mine for three years. The mine produced $33,000 before the company folded in 1890. The town of Hilltop began in 1906 when prospectors discovered gold ore very close to the original discoveries of the 1860s. Two mines, the Hilltop and the Independence, were soon producing ore. The Hilltop boom really got under way in 1908, and by 1909, close to 200 people were living at Hilltop. The town itself was well developed, with several fine houses, a school, numerous stores, and saloons. On February 17 a post office opened, with Chris Nelson serving as postmaster. Many solid wood-frame buildings were also constructed, and Hilltop gained a look of permanence. By 1912 mining activity was in full swing. The Hilltop Milling and Reduction Company (Frank Lebar, president) built a 10-stamp, 75-ton cyanide plant. The Philadelphia West Mine (formerly the Independence) produced $135,000 during a one-year period before the ore vein played out in 1913. Another prominent mine, the Kattenhorn, was discovered in 1912. The mine was owned by the Kattenhorn brothers, who shipped 100 tons of ore per month to a smelter in California. The ore assayed at 35 ounces of silver to the ton, and more than $95,000 in silver was produced before the mine closed in 1923. The Kimberly Consolidated Mines Company was also prominent at Hilltop. Incorporated in 1910, the company took over the holdings of the Philadelphia Western Mining Company, gaining control of 24 claims that showed values in gold, silver, copper, and lead. The ore from Kimberly Consolidated's mines was milled by the Hilltop Milling and Reduction Company until 1918 when Kimberly built its own 50-ton mill and cyanide plant. The main holding, the Kimberly Mine, ran out of ore in 1919, and the company folded after achieving a production of $161,000.

Hilltop's population varied greatly during the teens. By 1918, mining had slowed and only about 50 residents were left. The Hilltop Nevada Mining Company was organized in 1920 and purchased the holdings of the defunct Kimberly company at a sheriff's auction in March, but that move was not enough to save Hilltop. The start of the auto stage led to an exodus to Battle Mountain. Many of the fine houses built in Hilltop were moved there, and all of the businesses also moved. The last to go was the E. O. Swackhammer store, which moved in 1923. About 20 miners remained in the camp, though, as the Hilltop Nevada company expanded its operations. The company spent $300,000 in 1922 to build a 150-ton flotation mill, which began operations in October. The mill was not a success and operated only until September 1923. The mill was moved to Dayton (Lyon County) in 1934. The company continued to mine the district until the late 1920s. By the time operations were curtailed in 1927, $1.5 million had been produced. After Hilltop Nevada closed, Hilltop became a virtual ghost town. The school closed in 1927, and only 15 people still called the dying town home. The post office remained open, showing that Hilltop still had a small spark of life.

In 1930 the Hilltop Mine and a few additional claims were leased to the Buckingham Mines Corporation, which opened the mine in May and had minor production for a couple of years. A barite mine was put into production here during 1930 also. While not a great producer, it was active off and on until 1969. By the time it was abandoned for good, it had produced $300,000. Hilltop, however, didn't last that long. The post office closed on March 14, 1931. Hilltop still had about a dozen residents during the 1930s, but by 1940 the town had once again become a ghost. The remains at Hilltop are quite extensive. While only a few shaky wooden buildings and foundations mark the center of town, the number of mining and milling ruins is significant. Cement foundations of the Hilltop-Nevada Mill are located in a side canyon, below the townsite. The ruins of the Hilltop Milling and Reduction Company's mill are located on the side of the hill, behind the townsite. Many other mine ruins are scattered throughout Hilltop Canyon. If possible, plan to spend a day exploring Hilltop. Fresh spring water is available here, and the tree-filled Hilltop townsite makes for a relaxing visit. For pure pictorial and explorational enjoyment, Hilltop is one of the best spots in Lander County.

Hot Springs (Watts)

DIRECTIONS: *From Battle Mountain, head south on Nevada 305 for 35 miles. Exit right and follow for 0.25 mile to Hot Springs.*

Hot Springs was a stage stop during the 1870s. The stage station was located at the site of sixty hot springs. The camp took on additional importance in 1880 when the Nevada Central Railroad was completed and a ten-car siding along with a small railroad station was built here. Matt Smith ran the stop and started the Hot Springs Ranch. A wood frame house, which still stands, was built, as well as a barn, stone cabins, and corrals. A one-room schoolhouse was constructed in 1925, and children from Antelope Valley ranches came to school here until 1934. Hot Springs was abandoned in 1938 when the railroad folded and the rails were pulled up. Today original buildings still stand. The hot springs, which gave the site its name, have mostly gone dry.

The old station house at Hot Springs.

The remains at Hot Springs.

Jacobsville (Jacobs Spring) (Jacobs Well) (Jacobs Station) (Reese River Station)

DIRECTIONS: *From Austin, head west on U.S. 50 for 5.5 miles. Exit right at roadside rest area, which has a Jacobsville historical marker. Follow this poor road for 0.75 mile. Bear left and follow for 0.5 mile to Jacobsville.*

Jacobsville came into existence with the Pony Express. The station was named for Washington Jacobs, the district agent here. Indian trouble plagued the station, and it was burned to the ground in 1860. Sir Richard Burton visited the new station as it was nearing completion and reported:

> The station-house in the Reese River Valley had lately been evacuated by its proprietors and burnt down by the Indians: a new building of adobe was already assuming a comfortable shape. The food around it being poor and thin, our cattle were driven to the mountains. At night, probably by contrast with the torrid sun, the frost appeared colder than ever: we provided against it, however, by burrowing into the haystack, and despite the jackal-like cry of the coyote and the near tramping of the old white mare, we slept like tops.

The Overland Stage and Mail Company also used Jacobsville as a stop until the station was moved to Austin during that town's boom. Jacobsville was named Lander County seat in December 1862, when the county was carved out of Humboldt County, and the following year, an $8,400 courthouse was constructed. Jacobsville became the center for all miners and prospectors exploring the Reese River District. By the end of 1863, the town was at its peak, boasting a population of 400, two hotels, three stores, a telegraph office, and fifty homes. A post office opened on March 3, 1863, with James R. Jacobs as postmaster.

With the rapid growth of nearby Austin, Jacobsville's importance began to fade. An election in September 1863 was organized to select a county seat. Jacobsville could put up only feeble resistance, and the county seat moved to Austin. The post office closed on April 9, 1864, and the town was abandoned soon after that. A ranch that operated at the site for a while provided the only activity after the town folded. Today only scattered stone and brick foundations mark the site.

Kingston (Summit) (Morgan)

DIRECTIONS: *From Austin, head east on U.S. 50 for 11.5 miles. Then head south on Nevada 376 for 14 miles. Exit right and follow for 3 miles to Kingston.*

The Kingston townsite was laid out in 1864 as a result of the boom in nearby Bunker Hill. Kingston offered a more level area for building mills and houses. And since a couple of new mines were discovered near the mouth of Kingston Canyon, the new location was even more favorable for settlement. In 1865 the 20-stamp Sterling Mill was built. The following year, another 20-stamp mill was put into produciton. In June the New England and Nevada Consolidated Silver Mining Company graded a mill site just below the Sterling Mill. The old Colfax Mill (renamed the Smoky Mill) was moved here in July and started on September 24, working ore from the Mother Ledge Mine (Sante Fe District). Two other smaller mills also were active by 1866.

A post office opened on January 11, 1865, with Henry Schmidtlein as postmaster. The town of Kingston began to boom and within one year had a population of 125 and many businesses. Slowly, however, the mines faded and the mills began to run at less than capacity. The post office closed on December 26, 1867. In 1869 the Big Smoky Mill was moved to the booming White Pine District, near Hamilton. The district became dormant by 1870, and the clanging of mining machinery did not return to Kingston until the 1880s. Some mines were reopened in 1881 and the post office, now named Morgan, reopened on November 23. In August 1882 the Victorine Gold Mining Company gained control of the mines in the district, bringing some interest back to Kingston. A school was built in 1885 and remained open until 1923. Several business establishments returned. The activity was short-lived though, and all mining ceased the next year. The post office closed and only about 25 people remained.

The impressive ruins of the Sterling Mill.

The last revival began in 1906. The Kingston Mining Company gained control of the district's mines and enjoyed modest success for a few years. A 60-ton mill was built in 1909 and ran until 1911 when the company folded. Since then, no additional mining activity has taken place. The most impressive of the extensive remains at Kingston today are the imposing stone walls of the Sterling Mill. Other stone ruins remain in Kingston Canyon, but new houses have recently been built as part of a modern housing development and some of the ruins have been destroyed. The mill ruins alone make the trip worthwhile, however.

Lander (Bullion) (Campbell District)

DIRECTIONS: *From Tenabo, head north for 2 miles. Exit left and follow for 1.75 miles to Lander.*

Initial discoveries in the Bullion district took place in 1873. Two mines, the Campbell and Young and the Eagle, began producing ore during 1874. The mines closed down at the end of the same year, but produced $21,000 before they ceased working the site. The district remained silent until 1880, when a number of new mines were established and the camp of Lander formed. By 1883, Lander's population stood close to 100. Of the several mining companies actively engaged in the district, the Bullion Mining and Mill-

This lone brick wall is all that remains of the Lander City Bank.

ing Company was the most prominent. The company owned the Lady Don, Hooverdon, Mary Ann, Annie, Eagle, Lady of the Lake, Silverside, and Lovey mines. A 5-stamp mill was constructed during the spring of 1883. Forty men were employed by the company. The mill was started in May, processing ore from the Silverside Mine, and four bars of bullion, valued at $4,200, were shipped within the first month. In August a Bruckner furnace and 5 new stamps were added to the mill. A small company, the Nutmeg, also built a 5-stamp mill, the Frisbie, in 1882. The company was badly managed and in the summer of 1883 Bullion Mining acquired all of Nutmeg's holdings.

Both mills closed down in 1885, and mining activity gradually slowed. By 1890 only occasional efforts were made to mine the Bullion District. The camp wasn't abandoned, however; around 20 residents still lived in Lander during the 1890s. A school was built in 1889 and was used until 1898, serving students not only from Lander but also from nearby Utah Mine Camp and Mud Springs. Lander experienced a mining revival beginning in the summer of 1906, when mines, new and old, were worked. The Grey Eagle, Maysville, Bonnie Jean (formerly Lovey), Silver Prize, Silverside, and Silver Grey mines were active. A couple of mines were located southeast of Lander, and a rival townsite, Bullion, was platted on lower Indian Creek. A number of wooden buildings were quickly constructed, but the camp, which had a population of 25, folded within months of its organization. A post office opened at Lander on October 15, 1906, and Joaquin Bianchi was appointed postmaster. By 1907

Lander reached its peak, with 75 residents. However, the revival ended during the summer of 1909 when the mines ran out of paying ore. That fall the camp quickly emptied, and the post office closed on October 15. Lander had joined the ghosts by 1908. A couple of the mines were worked intermittently until 1921, and then the district was abandoned for good. Production from 1905 to 1921 was $625,000. Today stone ruins mark the Lander townsite. Only some faint wood rubble, half a mile east of Lander, shows that Bullion ever existed.

Ledlie

DIRECTIONS: *From Austin, take U.S. 50 west for 5.5 miles. Exit right at Jacobsville historical marker. Follow for 0.75 mile, then bear right and continue on for 1 mile to Ledlie.*

Ledlie was an important stop on the Nevada Central Railroad. The station was named after James Ledlie, one of the railroad's directors. Its location made Ledlie a natural shipping center as soon as the railroad arrived in 1880. As a result of demand for freight teams, as many as 250 mules were kept here. Supplies and mining equipment were shipped to budding mining camps like Ione, Grantsville, Jefferson, and Cloverdale via the Ledlie shipping yards. The huge volume of cargo inspired plans in late 1880 to construct a new railroad, the Ledlie and Cloverdale. An eighty-mile railroad bed was surveyed and portions were graded before the project was dropped in 1881. By 1890 the volume of freight had dropped drastically and Ledlie's importance faded. In 1906, however, the station was bustling once again; supplies for the booming camp of Skookum were shipped via Ledlie. Once Skookum folded in 1910, Ledlie was used only sparingly. When the Nevada Central Railroad folded for good in 1938, Ledlie ceased to exist. A town never really materialized here, since most workers lived in Austin. Today only a small collapsed wooden building and a solitary telegraph pole mark the site.

Lewis (Lower Town) (Middle Town) (Upper Town) (Dean)

DIRECTIONS: *From Galena Railroad, take good dirt road heading southeast and follow for 3.5 miles to Lower Lewis. Middle Town is located 0.5 mile farther on, and Upper Town is another 0.5 mile past that.*

Initial discoveries in Lewis Canyon were made in 1867 by two prospectors, Jonathan Green and E. J. George. These early discoveries didn't create any immediate excitement, and it wasn't until 1874 that Lewis began to form. Mines that began to produce ore during that summer included the Defiance, Henry Logan, Lousy Miner, Comstock, Keystone, Upland, Mountain Joe, and Miller mines. The Defiance was the most productive of these early mines. Two men named Fraser and Sweeney purchased the mine in April 1875 from Tom Hildreth for $40,000. The Eagle and the Star Grove, two major mines destined to be the best producers of the district, were discovered in 1875. As a result of this mining activity, a 10-stamp mill, the Eagle, was built by the Eagle Consolidated Mining Company in the fall of 1876.

In 1877 the Lewis townsite was laid out. Because Lewis Canyon was very narrow, three different sites were actually developed, appropriately named Lower Town (the main section of Lewis), Middle Town, and Upper Town. The first business in the district, Ramsdell's General Store, opened in 1875. Other Lower Town businesses included Len Pugh's Hotel, F. M. Sponagle's Drugstore, and Gentsel's General Store. Middle Town developed into a small town but never achieved the prominence of Lower Town. The businesses in Middle Town included the Felex Boardinghouse, Cozzens Meat Market, Hill Variety Store, and two saloons, Finley's and No. 1. The Miners' Union Hall was also located here. Upper Town, or Dean, was the smallest of the Lewis sections and was located near the Star Grove Mine. The Green boardinghouse and a saloon were built there. The camp was a company town of the Star Grove Mining Company.

A post office opened in Lower Lewis on April 15, 1878, with William Mills as postmaster. By 1880, the three sections of Lewis had a population of 216, and growth showed no signs of slowing. The Nevada Central Railroad was completed in 1880, and plans were made to build a spur line from Lewis Station, located three and a half miles away in Antelope Valley, to the mines in Lewis Canyon. This speculation led to the construction of two mills, the 40-stamp Highland Chief and the 15-stamp Star Grove. In addition, the Eagle Mill was enlarged to 15 stamps. The Highland Chief Mill and Mining Company (E. Young, president) began construction on its mill in April 1881. The mill, built by Beckett and McDowell of New York, consisted of 40 stamps, a 23,000-pound main driving wheel, and an 1,800-pound main belt and was equipped with Powell-White rotary furnaces. The total cost of the mill was $250,000.

The railroad spur, the Battle Mountain and Lewis Railway, was completed in April 1881 and greeted with great fanfare. The grading for the spur had begun on January 28, and the twelve-mile spur was finished by April 25. The engine was named the *John D. Hall,* in honor of the president of the railway. By summer, Lewis had a peak population of 700, its own railroad, and three large mills and was looking forward to the future. A newspaper, the *Lewis Weekly Herald,* began publication in town on November 23. Even the law came to Lewis: Abram Hull built a jail in Lower Town at a cost of $2,125. The droves of thirsty miners led to the construction of the Lewis Brewery, one of the few built in the state. The brewery was owned by two men named Cozzens and Fuchs, both of whom ran other businesses in the town. A school, which educated students until 1900, was built in 1881. However, not all was wine and roses for Lewis. A major labor dispute occurred at the Star Grove Mine during the summer of 1881. The company tried to cut wages and was forced to close when no miners would work. When the company tried to reopen the mine, intense violence broke out and three men were shot. One of them later died. A group of 250 Austin miners started out for Lewis to aid their mining brothers. When the company heard of their approach, the dispute was quickly resolved and the original wages were restored. The biggest blow to Lewis was a disastrous fire on December 5 that started in Sponagle's Drugstore and destroyed many businesses. That sad event signaled the beginning of the end for Lewis.

The Battle Mountain and Lewis Railway, in the face of mounting financial woes, was forced to curtail operations in early 1882. The Star Grove Mining Company went bankrupt after producing $300,000, and all properties sold at sheriff's auction in April. Another auction in May saw the Battle Mountain and Lewis Railway's railroad bed sold for $4,401 to H. D. Gates, a Lewis resident. In June, with no ore to process, the Eagle Mill also went bankrupt and shut down. By 1883 Lewis was rapidly becoming an abandoned town. The *Herald* folded on January 17, and the Star Grove Mill was torn down in February by G. W. Bothwell. The mill was dismantled and shipped, via Ledlie, to Bernice (Churchill County)—one of the few times the Battle Mountain and Lewis Railway was used after the Lewis boom went bust. The Highland Chief Mill closed down in 1883, sealing Lewis's fate. There was a slight revival during 1885 and the railroad was used for a short while, but a miners' strike closed the mines. The railroad was abandoned and the rails were torn up in 1890.

The post office moved to Dean in October 1894. By this time, only about 30 people lived in Lewis Canyon. The mines near Dean were operated intermittently until 1905, when all mining ceased. The post office closed on November 30, and soon only the ghosts of the past wandered through Lewis Canyon. An occasional leaseholder tried to make a go of it, but nothing substantial ever developed. The total production for the Lewis mines stands at just over $3.7 million.

As ghost towns go, Lewis provides an enjoyable visit for the traveler. In Lower Town, stone foundations and wooden rubble remain. The Lewis Cemetery is located to the north of the townsite. At the mouth of Lewis Canyon are stone walls of two mills, Star Grove to the left and Highland Chief to the right. Remains of Middle Town are located about half a mile into Lewis Canyon. Foundations of the Eagle Mill are on the right and stone foundations are on the left, among a huge stand of aspens. This is a beautiful spot to explore! The only standing buildings in Lewis Canyon, located at Dean, another half mile beyond Middle Town, are small cabins inhabited by miners working the Betty O'Neal Mine. Foundations and large tailing dumps remain. All in all, Lewis is definitely one of Lander County's best ghost towns, and offers something of interest for everyone. The visitor can, and should, spend a full day examining the different sections of town. Make time to visit Lewis. There are places for overnight camping in Lewis Canyon, with the best being near Middle Town. Fresh spring water is also available here.

McCoy (Wildhorse)

DIRECTIONS: From Battle Mountain, head south on Nevada 305 for 22 miles. Exit right and follow for 10.5 miles to McCoy.

McCoy took its name from James H. McCoy, who made the original gold discoveries in 1914. Only a few claims were staked and activity was limited. In 1918 the Quicksilver Mines Company worked the Ruby group. The company produced 78 flasks of mercury before leaving the district in 1925. It wasn't until 1928, after new gold deposits were discovered, that McCoy began to boom. Among these new gold mines were the Big Four, Gold Dome, Gold Pirate, Hancock (Mickey), and McCoy. The McCoy-Nevada Gold Mines Company, with H. D. Brown as president and James McCoy as vice president, controlled eleven claims in McCoy. The company was renamed the Nevada Gold Dome Mining Company in 1929 and in 1930 built a 20-ton amalgamation mill. The mill was constructed by Buckmaster, Mahoney, and Quick of Modesto, California.

A close-up of the McCoy operations in 1939. (USGS)

A camp of about 75 people formed at McCoy during this boom. Boardinghouses were built and a store opened. A school opened in 1929 and operated until 1940. In 1930, another company, the McCoy Consolidated Mines Company, entered the district. Its president, Charles Krengle, leased claims from Nevada Gold Dome in September. The gold deposits, however, went dry. By the end of 1931, virtually all mining activity had ceased in McCoy and the population dropped to a dozen during the rest of the decade. A small revival took place between 1938 and 1941. The Nevada United Gold Mining Company reopened the Dome Mine in December 1938, but only minor production occurred before the company folded in 1940. Nevada United Gold added $39,000 to the revival's production figures. The Wildhorse Quicksilver Mining Company produced 883 flasks of mercury between 1938 and 1941, making it the largest producer during those years. After mining activity ceased in 1941, McCoy was abandoned very quickly. The total production for the district was $95,000, with the bulk coming from the 1938 revival. Echo Bay Mining Company is currently operating an open-pit gold mine at McCoy.

Mound Springs

DIRECTIONS: *From Battle Mountain, head south on Nevada 305 for 25 miles to Mound Springs.*

Mound Springs was a siding for the Nevada Central Railroad beginning in 1880. A ten-car siding was constructed, and because of the abundant nearby springs, Mound Springs became an important water stop. A small house was built for the stationmaster, but that was the only development at the site. Mound Springs was used by the railroad until 1938, when the railroad folded and rails were pulled up. Barite was discovered in 1948, although production didn't begin until 1952. Owned by the F.M.C. Corporation's Inorganic Chemicals Division, Mound Springs Mine produced 500,000 tons of ore from 1952 to 1970. Today nothing marks the Mound Springs site except the barite mine open pit.

Mount Airy

DIRECTIONS: *From Austin, head west on U.S. 50 for 18 miles to Mount Airy.*

Some information has listed Mount Airy as a Pony Express station, but that is not the case. The station at Mount Airy did not come into existence until 1862—after the Pony Express had folded. Dry Wells, ten miles south, was the station site and also served the Overland Stage until the route was changed. The station was then moved to Mount Airy. The station, named for Sir George Biddell Airy (1801–92), an English astronomer, operated off and on until it was permanently closed in the 1890s. Today the faint crumbling walls of the station remain, as do a number of other stone foundations. One grave (a Mrs. Franklin, dead of smallpox in 1869) is also located at the site.

Mud Springs

DIRECTIONS: *From Tenabo, backtrack 1.5 miles to paved road. Exit left and follow for 4 miles. Exit left again and follow for 4 miles to Mud Springs.*

Initial discoveries in the Mud Springs District were made by Stephen Bedell in February 1885. His Rattler Mine produced $500-per-ton ore for a couple of months, but Bedell left the district in late 1885. Placer gold deposits were discovered in 1907 by Gus Fowler of Beowawe, leading to a small boom at Mud Springs. Other mines discovered between 1907 and 1908 included the Triumph (1908, 18 tons of ore yielded $60 per ton), Big Bug (two 100-foot tunnels), Bridal Wreath (30-foot shaft), Uncle Sam (50-foot shaft), and the Grey Eagle (located one mile west of the camp). By the summer of 1908, a camp of 30 had developed at Mud Springs. Most of the supplies for the camp came from nearby Lander. As a result, only one business, a saloon, opened.

The Grey Eagle Mine was the best producer of the district, with $25,000 between 1907 and 1908. A 250-foot shaft was dug, with lateral workings at the 60- , 115- , and 215-foot levels. By late 1908 the ore deposits in the Mud Springs mines began to run out. By spring 1909 all mining had ceased in the district, and by summer Mud Springs was abandoned. The only mining activity since occurred in October 1935. The A. O. Smith Corporation of Milwaukee sank a couple of shafts but found no substantial deposits and gave up after only a couple of months. Mud Springs has been a ghost ever since. The three buildings still standing at the site appear to have been cabins for the miners.

New Pass (Franklin)

DIRECTIONS: *From Austin, head west on U.S. 50 for 19 miles. Exit right onto poor dirt road. Follow for 2.5 miles. Keep left, ignoring right turn, and continue on for 4 miles to New Pass.*

Gold ore was discovered in late 1865 and the New Pass, or Franklin, Mining District was organized the following year. The name came from an old pioneer who believed that he had discovered a pass through the mountains, but he was mistaken. A number of mines were established during the spring of 1866, including the Gold Belt (original shipment yielded $1,871 per ton), Superior ($100 per ton), Oriental ($62 per ton), Ingoldsby, Lake, Central, and Churchill. The mines attracted attention during 1866 when Stetefeldt (designer of furnace fame) conducted an extensive review of the New Pass District. He calculated that it would be profitable to build a 20-stamp mill here and estimated that it would cost $62,500, plus another $11,500 to prepare the mines for consistent production. These plans never materialized. The old 5-stamp Ware Mill was moved to New Pass from Austin in October 1867. New Pass peaked during 1868 when a camp of 50 formed. Buildings were erected and a mercantile store and saloon opened for business. The mines declined during the late 1860s, and the New Pass Mill closed in 1871.

New Pass was abandoned until fall 1889. Two companies, the Nantucket Mining Company and Starrett, Raindoh and Company, worked the Rabbit Foot, Sheridan, and Yellow Bull mines. Both gave up by summer 1890. A revival from 1900 to 1904 saw 50 residents return to New Pass. A post office named Franklin, with Emma Bonner as postmistress, opened on February 20, 1900. The name was changed to New Pass on May 2, and Charles Littrell was named postmaster. The mining activity ended in 1903 without any substantial production. The post office closed on February 28, and by summer the camp was empty. The Nevada Austin Mines Company worked the district during

The New Pass Mine.

World War I. A 75-ton cyanide mill was constructed to treat the company's copper ore, but the company folded in early 1919 and the property was sold to pay a $30,000 mortgage.

Intermittent activity took place at New Pass from 1929 to 1941. New Pass Gold Mines, Inc. (Parker Liddell, president) established a camp in 1929. A 30-mile road to Campbell Creek Ranch, near Carroll Station, was built, and a 50-ton Hardinge crushing mill was constructed in 1930. The company worked the district until 1937. In 1939 Wayne Smith and Howard Snyder began to work the Thomas W. Mine and $1,840 was produced during the next two years. Mining activity has been periodically attempted since then, but nothing of substance has actually been produced. Today only a couple of wooden buildings remain. In addition, stone ruins and remains of three mills are scattered around the New Pass site.

Pittsburgh (Pittsburg)

DIRECTIONS: *From Lander, continue west for 11 miles to Pittsburgh.*

Initial discoveries were made in Crum and Maysville canyons in 1878. The Morning Star Mine was located on May 12 by Thomas Morgan, who had been living in Lewis since 1875. Morgan organized the Morning Star Gold Mining Company with partners John Cardwell and a Dr. Bean. In 1880, Morgan located two additional mines: Lady Carrie (named after his wife) and the Pittsburgh (soon to become the best producer of the district). In 1884 Morgan built a 20-ton, 10-stamp mill. On May 15, 1885, he gained complete control of Morning Star Gold Mining. In January 1886, Morgan sold all of the company's holdings to an English company, Thurber, Gates, and Mooring, for $400,000. The company was reorganized and Pittsburgh Consolidated Gold Mines, Limited, was born. Besides the Pittsburgh, Morning Star, and Lady Carrie mines, the company also owned the Cumberland, Evening Star, and Ida Henrietta mines. In 1887 the company built a $60,000, 100-ton, 20-stamp mill, which replaced Morgan's old mill. The company printed 80,000 one-pound shares to help raise money for the mill's construction.

The camp of Pittsburgh began to grow during the mid-1880s, and a townsite was laid out in Crum Canyon (next to the mill), at the mouth of Hilltop Canyon. The mines were located about a quarter mile up Hilltop Canyon, and a tramway to the mill was constructed. Between 1886 and 1887, $126,000 in gold was produced, and Pittsburgh soon had a population of 100. Around thirty buildings were crowded into Crum Canyon. A couple of stores and saloons opened their doors. A post office opened on October 11, 1888, with William J. Black as postmaster, and operated until August 15, 1893. It is interesting to note that on March 5, 1892, the post office changed its name from

Pittsburgh to Pittsburg. Pittsburgh Consolidated Gold Mines folded in 1891, but the property was immediately purchased by W. E. Dean. He built a 10-stamp mill near the camp of Dean, located in upper Lewis Canyon. The mill started production in December 1892, employing 30 men. However, production dropped dramatically and Pittsburg's population soon fell to about 20, all miners. All the businesses left, and the miners had to trek up to Dean for supplies and a drink. Enough people still called Pittsburg home to have the post office reopened on December 6, 1897, and it operated until May 15, 1900. The mines and mill were finally abandoned in 1906 after a few years of progressively poorer production. At the time of abandonment, the Morning Star had two miles of workings, while the Pittsburg had 4,500 feet. Pittsburg was totally empty by 1908, and no revivals ever took place. Today only one old shack still stands. The only other ruins are the large foundations of the 20-stamp mill. Mine ruins are located a quarter mile farther into Hilltop Canyon.

Ravenswood (Shoshone)

DIRECTIONS: From Silver Creek, take poor road heading west for 6 miles to Ravenswood.

A group of Austin prospectors discovered silver at the edge of the Shoshone Mountains in 1863, and a small mining camp sprang up at the base of Ravenswood Peak. In 1865 the district was described as "one of the most important mining localities in the whole Reese River region." Springs at the camp provided the mines with five barrels of water per day. The United States Gold and Silver Mining Company bought all the district's mines, including the AJC No. 5 (formerly the Shoshone), Pine, Queen, Lombardy, and Red Bird. The district never showed any of the promise exhibited by early propaganda though, and the camp folded in 1870 after producing $10,000.

Some new activity did take place in the early 1880s and 1890s, but not until 1906 did serious exploration get under way. Tasker Oddie, of Tonopah fame, gained control of the district's properties and instigated a small boom that brought 30 to 40 miners to Ravenswood. The Queen Mine was expanded to a depth of 240 feet, with two 120-foot drift tunnels. Unfortunately, the ore deposits continued to be small, and by 1908 the veins had disappeared. The camp was abandoned and no activity has taken place since. Nothing but scattered rubble marks the site today.

Ravenswood was also the name of a siding on the Nevada Central Railroad. The siding was located three miles north of Vaughn's on Nevada 376. Nothing remains at that site either.

Only scattered railroad ties mark the Silver Creek stop.

Silver Creek

DIRECTIONS: From Austin, take U.S. 50 west for 0.6 miles. Exit right on Nevada 305 and follow for 16.3 miles. Exit left and follow poor road for 1.75 miles to Silver Creek.

Silver Creek was a stop on the Nevada Central Railroad from 1880 until 1938. A forty-car siding was constructed and a small settlement formed. A school, built in 1881, was used until 1914. Chrissie Watt served as schoolteacher for many years. Nearby mining claims were worked in the 1890s by Professor E. Craine, a teacher at Midas (Elko County). He found boulders that were laced with silver, and the first shipment yielded $4,800 per ton. However, only a couple of these boulders existed. A few collapsed wooden buildings mark the site today. Hundreds of railroad ties, still embedded in the railroad right-of-way, mark the Nevada Central Railroad. Visitors should beware of scorpions when exploring this site.

Simpson Park

DIRECTIONS: From Austin, take U.S. 50 east for 5.7 miles. Exit left onto Nevada 306 and follow for 4 miles to the Willow Creek Ranch. Exit right and follow rough road for 1.5 miles to Simpson Park station.

Simpson Park, the site of a Pony Express station, was named for Captain James Simpson. The station was built in spring 1860 but on May 20 it was burned by Indians, who also killed the stationmaster, James Alcott, and stole all the stock. When Sir Richard Burton visited the site on October 12, he reported that the station was being rebuilt. After the Pony Express folded, Simpson Park was used by the Overland Stage and Mail Company until the route was moved south to Austin. After that, a ranch operated here for many years. A school was built at nearby Willow Creek Ranch and was active from 1927 until 1941. Today huge stone foundations of three buildings mark the Simpson Park site. One is easily identified as the station house and another as a horse barn and blacksmith shop. The third is of unknown origin. To the south of the foundations are remains of the old station corral, and within it are the remains of a small stone building. A couple of other small stone ruins are scattered near the site. All of these foundations are located in a private field, so visitors should ask permission before venturing onto the site. A cemetery is located on the other side of the road, just north of the site, on a small hill. The six graves, including Alcott's, are almost indiscernible from the surrounding landscape. As late as the 1950s, grave markers were visible, but now only a very sharp eye can pick out the lonely graves of these nameless pioneers.

These faint foundations are all that remain of the Simpson Park Pony Express station.

The short-lived town of Skookum had an air of permanence in this 1908 photo. (Nevada Historical Society)

Skookum

DIRECTIONS: *Located 1 mile east of Gweenah.*

An Indian discovered some rich gold and silver float in early 1906. He sold his find to the Lemaire brothers of Battle Mountain. The camp of Skookum soon formed, and a small boom occurred during the summer of 1908. Skookum, and its sister camp of Gweenah, had a population of 200. Stores and saloons opened at Skookum, and a short-lived newspaper, the *Times,* began publication. By 1909, the boom was over, as ore values dropped dramatically. Some minor production continued until 1914, when the district was abandoned. A small amount of production also took place during the summer of 1926, when Virgil Clark and George Govich worked the Greenough Mine. Total production for the district was under $15,000. Today only a collapsed shack and a mine shaft mark the site.

Smith Creek Station (Maestretti) (Edwards Creek)

DIRECTIONS: *From Brown's Station, continue west on Nevada 2 for 6.4 miles. Exit right and follow for 9.5 miles. Exit left and follow for 3.5 miles to Smith Creek.*

Smith Creek Station was the first Pony Express stop in Shoshone country. The station was named for one of Simpson's men, Captain Smith, when the Pony Express survey party passed the site on May 30, 1859. Smith

Smith Creek Ranch, around the turn of the century. (Nevada Historical Society)

Creek had quite a violent history. The *Territorial Enterprise* reported in August 1860:

> One day last week H. Trumbo, stationkeeper at Smith Creek, got into a difficulty with Montgomery Maze, one of the Pony Express riders, during which Trumbo snapped a pistol at Maze several times. The next day, the fracas was renewed when Maze shot Trumbo with a rifle, the ball entering a little above the hip and inflicting a dangerous wound. Maze has since arrived at this place (Carson City) bringing with him a certificate signed by various parties, exonerating him from blame in the affair and setting forth that Trumbo had provoked the attack.

In another incident, a Pony Express rider, William Carr, killed a man named Bernard Chessy, with whom he had quarreled earlier at Smith Creek. Carr later achieved the dubious honor of being the first man legally hanged in Nevada Territory. Yet another Pony Express rider, Bart Riles, met misfortune at Smith Creek; he was killed in a riding accident on May 30, 1860.

Sir Richard Burton visited the station on October 14, 1860, and described the site:

> The station was sighted [*sic*] in a deep hollow. It had a good stone corral and the visual haystack, which fires on the hilltops seem to menace. Amongst the station folks we found two New Yorkers, a Belfast man, and a tawny Mexican named Anton. The house was unusually neat, and displayed even signs of decoration in the adornment of the bunks with osier taken from the neighboring creek. We are now in the land of the Pa Yua (Paiute) . . . I observed however, that none of the natives were allowed to enter the station house.

After both the Pony Express and the Overland Stage lines folded, Smith Creek became a successful ranching operation. The Maestretti family ran the ranch for three generations. A post office by that name opened on July 5, 1904, with Antonio Maestretti as postmaster. The office served the ranch until July 14, 1906. A school was established in 1903 and was used off and on until 1941. The ranch is still active today, and some of the original buildings remain. The adobe section of a half-adobe, half-rock building is part of the original Pony Express station. Other old willow- and thatched-roof buildings also remain.

The Smith Creek schoolhouse. (Nevada State Museum)

Starr

DIRECTIONS: From Galena (Railroad), continue south on Jenkins road for 2.75 miles. Exit left and follow rough road for 2.5 miles to Starr.

Starr was a small and short-lived mining camp that sprang up during the summer of 1881. The camp was named for A. M. Starr, the co-owner with Sam Groves of the Starr and Grove Mine (not to be confused with the mine of the same name in nearby Lewis). A small camp of about 25 formed near the mine, and a post office, with Jonathan Green as postmaster, opened on October 18. The promise that Starr showed in 1881 quickly faded in 1882. By spring, the Starr and Grove Mine was mined out. The post office closed on May 12, and by summer, the camp had been totally abandoned. Mine dumps and scattered rubble mark the site.

Telluride

DIRECTIONS: From Copper Canyon, head south for 0.75 mile. Exit right and follow this road for 1 mile. Bear right and continue for 3 miles to Telluride.

Though the camp of Telluride didn't form until 1911, mining activity had taken place at the site in 1875. During that summer, the Battle Mountain Mining Company built a 30-ton concentrator, which processed ore from several Copper Canyon mines. The mill closed down in 1876, and the district was empty until 1910. In November, J. Hutchins discovered the Dollar Mine. He sold his find to Thomas Kearns, who organized the Dollar Mining Company. The mine was at an altitude of 7,500 feet, and its ore assayed as high as $500 per ton. A townsite was laid out in June and soon Telluride had a population of 50. Small buildings were put up, and a store and saloon opened. However, the distance and difficulty of access prevented the camp from further development. The mine was abandoned in 1912 and by summer Telluride had joined the ghosts. Only scattered rubble now marks the site.

Tenabo (Raleigh)

DIRECTIONS: From Beowawe, take Nevada 306 south for 21.5 miles. Exit right and follow for 1.5 miles to Tenabo.

Silver and gold ore was discovered during the summer of 1906, and immediately a rush developed. A post office opened on December 7 and Hiram Mills was appointed postmaster. A townsite was platted just east of the mines, located in Mill Gulch. By 1907, Tenabo had become a large town, boasting a population of 1,000. The residents supported a wide range of businesses: restaurants, hotels, assay office, grocery store, lumberyard, school, and many popular saloons. A triweekly stage to Beowawe was set up. A mill was built in nearby Mill Gulch. The largest mine in the district was the Little Gem Mine. The mine was discovered in 1907 and soon had a 400-foot shaft and more than 900 feet of drift work. The district's other mines included the Phoenix (two-compartment, 250-foot shaft), Gold Quartz (308-foot shaft), and Two Widows mines (110-foot incline shaft).

By 1909 most of Tenabo's mines were controlled by the Tenabo Mining and Smelting Company. The company, with W. Mont Ferry as president, bought out the holdings of the Gem Consolidated Mining Company and Reliance Mining and Milling Company. In addition, three more mines were purchased: Little Gem, Copper Hill, and Two Widows. By 1911, the expense of producing ore became higher than the value of the ore. The post office closed on July 31, 1912, and Tenabo quickly declined. In 1916 a new camp formed in Mill Gulch. A. E. Raleigh discovered placer gold there, and a camp by the same name sprang up. Water for the "rockers" was brought to Raleigh from Indian Springs. The camp didn't last very long, although placer mining continued. In the 1930s, a huge floating dredge operated for a while, producing significant amounts of gold. The Mill Gulch Placer Mining Company bought the Raleigh property in 1936, and a dragline dredge and washing plant operated from May 1, 1937, until April 3, 1939. In 1938 Mill Gulch had the distinction of being the top placer producer in the state.

The Tenabo Mining and Smelting Company sold all of its holdings in 1920, after producing 12 million ounces of silver and 25,000 ounces of gold, to the Tenabo Consolidated Mines Company. Tenabo added those holdings to its own mines, the Tenabo and Gold Quartz. The company did not enjoy any success and folded in the early 1930s. A couple of additional dredging companies, Idaho-Canadian Dredging and Yreka Gold Dredging, worked the district during the mid-1940s. The Tenabo District was pretty much abandoned until the Mid-West Oil Corporation gained control of the Tenabo mines in 1972. In 1975, Mid-West sold out to the Tenabo Gold Placers Limited Partnership. Today the property is actively being worked by the Flowery Gold Mines Company of Nevada. People still live in the camp, but most miners live in nearby Crescent Valley. Wooden buildings stand at the townsite. In Mill Gulch more buildings and mill ruins remain.

Trenton (Trenton Mill) (Buffalo Valley District)

DIRECTIONS: From Battle Mountain, head west on U.S. 80 for 14.5 miles. Exit south at Valmy interchange and follow this poor dirt road for 2.5 miles. At fork, bear right and continue for another 2.5 miles. At this fork, bear right and continue for 2 miles. Bear left and continue for 1 mile. Take left and follow deteriorating road for 2.5 miles. Exit left and follow for 2 miles to Trenton mine and mill site.

Silver was discovered at the edge of Buffalo Valley in June 1869. The Trenton Mine was owned by Lott and Company. The mine consisted of three tunnels (400 feet, 200 feet, 100 feet) and two shafts (55 feet, 40 feet). A small 5-stamp mill was built in 1870, and the ore assayed at $157 per ton. Fifteen people lived at the remote camp and supplies were brought in from Battle Mountain. The mine and mill closed in October 1873 after producing $23,000. The district was revived in 1912 when Clyde Garser located the Buffalo Valley Mine. The Buffalo Valley Mines Company was organized, but the first ore shipment wasn't made until 1916. In 1924, a small, 10-ton cyanide mill was constructed. The mine was active until 1940, but very little ore was actually produced. Then, in 1940, Manganese was discovered. The Black Rock Manganese Mine produced $300,000 before closing in 1947. A small mill was built during this period, but it was a failure. Today mine dumps and mill ruins remain in Trenton Canyon.

Vaughn's (Clark's)

DIRECTIONS: From Austin, take U.S. 50 west for 0.6 miles. Exit right onto Nevada 305 and follow for 24.1 miles to Vaughn's.

Vaughn's was a small railroad stop and siding on the Nevada Central Railroad from 1880 until 1938. The station was named for L. S. Vaughn, a nearby rancher. Besides the small freight platforms used to ship outgoing produce, only a thirteen-car siding was constructed. Nothing substantial was ever built or remains to mark the site.

Walters

DIRECTIONS: From Vaughn's, continue north on Nevada 305 for 5.5 miles to Walters.

Walters was both a stop and a recreational spot on the Nevada Central Railroad. The station and siding were active from 1880 to 1938. The station was named for William Walters, a prominent nearby rancher. Walters was the site of the "Austin Picnic Train," which brought weekend picnickers

The solid Walters station still stands despite heavy vandalism and graffiti.

to enjoy a relaxing day in the grassy areas near the station. A baseball diamond was laid out, and the rapidly growing sport of baseball became part of the weekend festivities. Today the solid stone station still stands, one of the better remaining station houses on the old railroad. Unfortunately, the stately building has been the victim of graffiti vandals.

Yankee Blade (Yandleville)

DIRECTIONS: *From Austin, head west on U.S. 50 for 0.6 miles. Exit right on Nevada 305 and follow for 1 mile. Exit right and follow for 3.5 miles to Yankee Blade. Road is extremely rough and four-wheel transportation is recommended.*

Yankee Blade was one of the many Reese River towns that sprang up during 1863. The camp was named after a New England newspaper, the *Yankee Blade*. The Whitlatch Yankee Blade Mine was the first prominent producer. The mine was located in June and produced a considerable amount of

$400-per-ton ore during the next few months. By April 1864, thirty cabins, most made of stone, were scattered near the district's mines. One prominent early mining company was the Confidence Mining Company, which was formed in late 1864 and consolidated the mines owned by B. C. Hill, L. H. Newton, W. H. Fearing, Charles Boyden, and H. Hazeltine. The mines included the Confidence, Maid of Athens, St. Jo, Revenue, and Lydie Allen. In spring 1865 construction began on the Confidence Mill. The 15-stamp mill was built by Nevins and Hussy and was fired in June. By 1865 Yankee Blade was a booming mining camp. A number of mills were built, and scores of new mines began to produce ore. The 20-stamp Keystone Mill was the most impressive of the new mills. The mill cost $125,000, and another $50,000 was spent on other company buildings. Workings of the mill included four batteries of 5 stamps each, a 65-horsepower engine, and four reverberatory furnaces. The mill, however, had a short life. The structure and a number of nearby buildings were destroyed by fire on April 6, 1868. Another prominent mill, the Mettacome, was built and owned by the Lane and Fuller Company. The mill, which had ten 900-pound stamps and four roasting furnaces, treated ore from a new and very rich mine, the Buel North Star. During 1866 three other mills were built: the 10-stamp Empire (started in May, later enlarged to 15 stamps), the 15-stamp Midas, and the 8-stamp Butte.

Among the new discoveries in Yankee Blade were the Yankee Blade (150-foot shaft, 500-foot crosscut tunnel), Midas Flat/Peerless ($440 per ton), Patriot (700-foot shaft, $30,000 produced), Morris and Cable ($218,000 produced), Chase ($22,000 produced), Monroe ($256 per ton), and Cicero mines.

In 1867 water was struck in the mines, closing down most of the district. By the end of 1867 only the Mettacome and Butte mills were still operating. The Confidence Mill was sold in February 1867 to the La Plata Silver Mining Company and moved to Park Canyon (Nye County). In August 1872 a partial revival took place. The Kelly and Kling Mining Company, employer of 20, worked several claims. The Mettacome Mill was taken over by the Pacific Mining Company, Limited. Only about 50 people still lived in Yankee Blade by 1875, but that turned out to be the camp's largest production year, with $42,000. The bulk of ore came from a new mine, the True Blue, and three older mines, the Patriot, Morris and Cable, and Chase. But it was Yankee Blade's last gasp. Only the Morris and Cable and Chase mines continued producing limited amounts of ore into the 1880s. After that, only an occasional leaseholder worked the district until the turn of the century.

A small revival began when a few of the mines were reopened in 1900 and worked until 1911. In 1911, the Maricopa Mines Company bought and worked the Watt, Morris and Cable, Chase, Patriot, and True Blue mines. The company built a small $6,000 mill in 1913, but the revival faded in 1914 and by that fall the district was once again abandoned. The last activity in Yankee Blade took place in 1937 when the Austin Syndicate reopened the True Blue Mine, but without any success. Today a visit reveals a multitude of picturesque stone ruins scattered throughout the townsite. In addition, walls of three mills remain at Yankee Blade. The site is very difficult to reach, but the effort is well rewarded.

A Short History of White Pine County

White Pine county was created out of Lander County on April 1, 1869, as a result of the "White Pine Rush" to Hamilton and famed Treasure Hill. Thousands came from all over the West and made Hamilton one of the largest towns in the West. The rush faded during the 1870s, but new discoveries at Cherry Creek and Ward kept White Pine County high on Nevada's production lists.

The Pony Express had numerous stations in White Pine County. The Indians resented this intrusion and burned some of the stations. A fort was established at Schellbourne until problems subsided. With the many scattered mining camps, freight and stage lines crisscrossed the county.

Many towns developed, prospered, and died. Each achieved a small bit of prominence before disappearing. In the early 1900s, however, copper took over. The huge open pits at and around Ruth were among the largest in the country. For many years the copper mines were the lifeline of White Pine County. The Nevada Northern Railway was completed in 1906 and served the mines and people of White Pine until it recently closed. The low prices of copper and expensive repairs required at the McGill Smelter forced the smelter to close. More than $1 billion had been produced by White Pine County mines.

But White Pine has rebounded from adversity. Microscopic gold has proved to be a new bonanza here, as well as in other counties in Nevada. Current operations at Alligator Ridge, Bald Mountain, Green Springs, Robinson District, Ward, and Taylor are producing over a million ounces of gold a year. Mining has returned to stay in White Pine County.

In addition, White Pine County contains the country's newest national park, Great Basin National Park. Also near the park are the fantastic Lehman Caves, with spectacular formations and rugged beauty that attract visitors from all over the West. At the East Ely Depot, Ely offers the "ghost train," a ride up the canyon to Ruth that turns back the clock to a bygone era. The surge in the tourist trade has been a boon to Ely and White Pine County.

With the mining revival, new industries moving in, and a new state prison, White Pine County has fought its way back from hard times and is looking toward new horizons and prosperous years ahead.

White Pine County

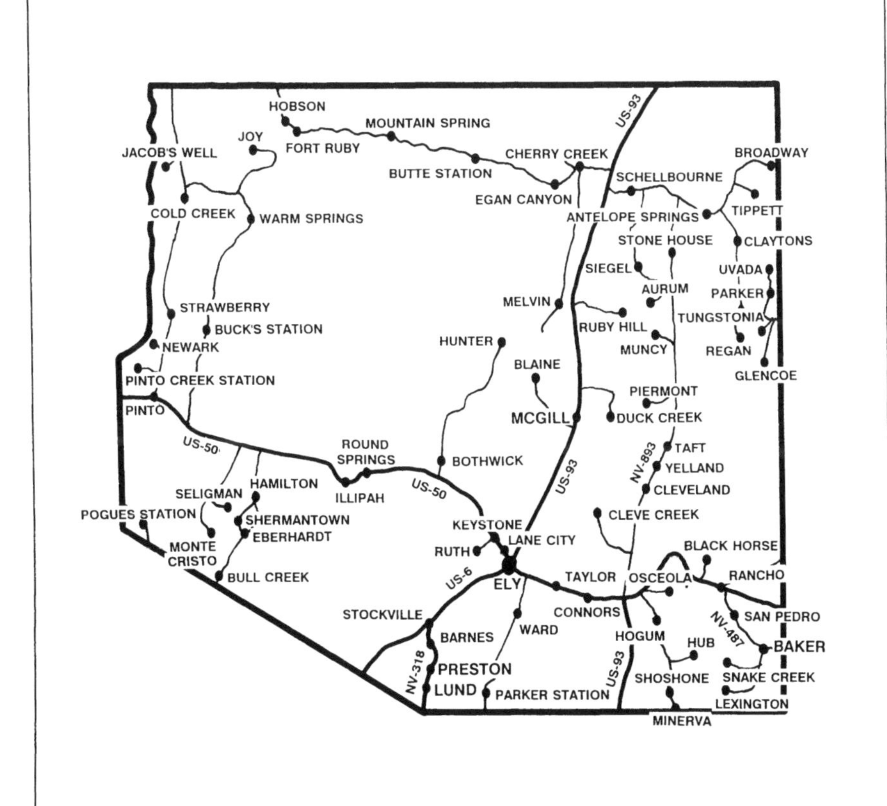

The only building remaining at Antelope Springs. (Nevada State Museum)

Antelope Springs

DIRECTIONS: From Tippett, head south on the Lincoln highway for 0.75 mile. Exit right and follow for 2.5 miles to Antelope Springs.

Antelope Springs was a Pony Express stop beginning in 1860. The station was burned by Goshute Indians in June and wasn't rebuilt until the Overland Stage and Mail Company took over the line. The company built a small log station, repaired the Pony Express corrals, and used the site until 1869. Antelope Springs was part of the Hamilton–Elko line. Flora Bender, who visited Antelope Springs on July 13, 1863, reported, "There is a small stage station and nice spring here, but no grass." Only a small log building remains at the site. There are differing opinions as to whether that building is the station or one that was located inside the corrals at Antelope Springs.

Aurum (Silver Canyon) (Doughberg)

DIRECTIONS: From Schellbourne service station on U.S. 93, exit on Nevada 2 for 14.8 miles to stone house. Exit Nevada 2 and head south for 3.5 miles. Exit right and follow this rough road for 3 miles to Aurum.

Initial discoveries were made in the Silver Canyon District in April 1869 by two men named Chisolm and Ramsdell. Most of the mines were controlled by the Grace Mining Company (James Gamble, president), whose holdings included the Grace, National Whiting, Marian No. 1, Nellie and Albert, and Dictator mines. The small camp of 50 reached its peak during 1872, and by 1873 Silver Canyon was empty. In 1878 new discoveries made by Dr. Brooks and A. Lawler included the Sadie L., Bluebell, and Buckhorn mines. A new town, called Aurum, sprang up. Brooks contracted to have a 10-stamp mill constructed by Rankin, Brayton, and Company of Cherry Creek. The stamps for the mill were brought from an old mill in Ward. The Aurum Mill was located at the mouth of Silver Canyon and started on January 1, 1881. Twelve men were employed at the mill and another 18 in the mines. Ore for the mill was transported, via tramway, from the Blue Bell Mine.

Aurum had become a fair-sized town by summer 1881. A post office, with John Robertson as postmaster, opened on April 4. A store (Garraghan and Poujade), a saloon (James McNulty's), a blacksmith shop, and two boardinghouses opened for business. A school, with Mollie Gripper as schoolmistress, opened on November 21. By 1882, however, Aurum experienced a mining slowdown. The mill shut down, and by July only 8 men were working Aurum mines. The post office left Aurum, but it retained the name and drifted all over Spring Valley during the following years. The office operated at Piermont, Muncy, Taft, and a couple of other places before finally folding in 1938. Disaster struck Aurum on February 11, 1884, when a snowslide leveled the Sadie L. Mine and destroyed the mine's boardinghouse. Three men were killed: H. W. Mickel (foreman of the mine), John Fox, and Wallace McCrimmon.

Aurum experienced a revival beginning in 1887. The Aurum Mine, showing values of $300 per ton, was discovered by Simon Davis and George Palmerton. In July, the two sold the mine to a Salt Lake group for $35,000. By 1888, Aurum once again had more than 50 residents. Freight was brought in from Toano (Elko County) at a cost of $85 per ton. Stage lines were set up, running to Cherry Creek, Osceola, and Ibapah, Utah (Deep Creek Station). Ben Sanford, a store owner and Aurum postmaster, and his partner Simon Davis shipped 300 tons of high-grade silver ore to Wells in 1889. The revival peaked in 1897 and 1898, when Davis and Albert Erickson discovered manganese silver ore. The ore was sent to smelters in Salt Lake City, where it returned $400 to $500 per ton. The active mines in the Aurum District during this period were the Clara, Florence, Iron Duke, Lucky Deposit, and Black Eagle (the richest of the new mines).

By 1906 the district was abandoned except for Simon Davis, who continued to prospect along Silver Canyon. Davis organized the Lucky Deposit Mining Company in 1914 and gave Aurum its last gasp at survival. The following year the White Pine Copper Company (John F. Cowan, president) bought ten claims in Aurum. Developments included 1,500 feet of workings, and the ore showed 5 percent copper and 40 ounces of silver to the ton. The company expended $50,000 in 1918 and 1919 to develop the holdings, but that investment proved to be the company's undoing and it folded in 1920. Soon Simon

Davis was once again the only resident in Aurum. When he left in the mid-1920s, Aurum permanently joined the White Pine County ghost roll. Today foundations and ruins of the mill remain at the mouth of Silver Canyon. A cemetery is located on the small hill that overlooks Aurum. Some recent exploration has taken place, and it is possible that new mining activity might begin in earnest soon.

Babylon

DIRECTIONS: *From Hamilton, head north for 4 miles, bearing left at 0.5 mile. Exit left and follow for 1.25 mile. Exit left again and follow for 2 miles to Babylon.*

Babylon was one of many short-lived camps that sprang up in the hills around Hamilton during the late 1860s. The camp formed in summer 1868 after the Conover Brothers Mine was discovered. A townsite was platted in March 1869. Babylon was an extremely cold camp because of its high altitude, and the miners were forced to keep fires going even in the middle of summer. Only a handful of prospectors ever lived here, and nothing more substantial than dugouts was ever constructed. Babylon was abandoned in the fall of 1869, and the *White Pine News* reported cynically, "Babylon has Fallen!" Today only sunken holes mark the sites of the camp's few buildings.

Barnes

DIRECTIONS: *From Preston, head north on Nevada 38 for 4 miles to Barnes.*

Barnes was the name of a post office that served ranches in lower Jakes Valley from July 28, 1902, to February 28, 1907. The post office was located on the Judd Ranch, and owner Joseph Judd acted as postmaster. Today only scattered rubble marks the site.

Black Horse

DIRECTIONS: *From Baker, head north on Nevada 487 for 5.3 miles. At junction with U.S. 50, head east on U.S. 50 for 7 miles. Exit right and follow for 2 miles. Bear left and follow for 0.5 mile to Black Horse.*

Black Horse was discovered entirely by accident. In March 1906 a prospector from Osceola, Tommy Watkins, sought shelter under an overhanging ledge during a spring storm. He chipped off some samples, and they proved to have a very high gold content. On March 6 Watkins—and 99 others—staked claims, and Black Horse began a tremendous boom. By April the camp had a population of 400 and businesses operating that included three stores, three saloons (one of which was named the Bucket of Blood), two boardinghouses, a blacksmith shop, and a barbershop. Most of these early businesses were initially housed in tents. A school was built, and a post office opened on September 17, with J. H. Mahigan as postmaster.

Some of the richest ore ever discovered in Nevada came from Black Horse. A sample from the Mabel group, owned by F. C. McFall and Fred Schrott, returned more than $100,000 per ton. Rich ore, valued at $50,000 per ton, was also removed from the Black Horse Mine. Other mines in the district included the San Pedro (sold for $75,000 in 1907), Grasshopper (owned by Lon Heath), California (Mariott Brothers), Red Chief (Hamilton Brothers), Lucky Boy, Buchanan, and Campbell. The largest mine was the San Pedro, covering 21 claims. The ore of the Black Horse, while incredibly rich, was located in very small veins. One moment the ore was there, the next it was gone. In January 1908 owners of the Mabel Mine built a crude mill in nearby Willow Patch. The following year, the owners of the San Pedro Mine built a $55,000, 20-stamp mill. The Amalgamated Nevada Mines Company gained control of most of Black Horse's mines in 1910, but the richest ore had already been removed. Only $19,000 was produced during the next two years.

Ore in the mines ran out in 1913, and Black Horse quickly became a ghost town after producing close to $1 million during its short existence. The post office closed on March 24, 1914, and there was no activity until 1933, when the Pauline, or Bellander, Mine was discovered and worked intermittently until the late 1940s. In 1942 another mine, the Gold King, was discovered. A 25-ton cyanide plant was built the next year to treat tungsten ore from the Gold King Mine. This activity lasted until 1954, but the town itself never revived. From 1933 to 1954, another $108,000 was produced. Today most of Black Horse has disappeared. No buildings are left, and only rubble and faint foundations remain. There are many large dumps that lend identity to the crumbled buildings. Open mine shafts are scattered throughout the townsite, and visitors should be very careful while exploring.

Blaine (Granite) (Campbell's Ranch)

DIRECTIONS: From McGill, head south on U.S. 50 for 3.3 miles. Exit right and follow for 15 miles to Blaine.

Initial discoveries were made on the eastern slope of the Egan Range near Campbell's Ranch in 1878. However, production was minor and the district didn't really become active until 1894. W. D. Campbell, owner of the nearby ranch, discovered the Blaine Mine (and named it in honor of James G. Blaine, a presidential candidate). A 5-stamp mill was bought by Campbell and moved from Telegraph Canyon to Blaine. Other mines opened included the Stinson (Jack Stinson, discoverer, $80 per ton), Alvin, and Ben Hur. A small camp formed, but ore values were too low to maintain production. By 1896 the camp had been abandoned, and only Campbell continued to work the mines.

A revival began in 1900. The mill was repaired and renamed the Ben Hur. In 1902 two mines were discovered: the Bunker Hill Sullivan (discovered by Fred Francis) and Cuba Lead (worked intermittently until 1960). Blaine really began to boom in early 1908. Governor Denver Dickerson bought the Blaine Silver Mine in March, and the resulting promotion led to a rush of a few hundred. A townsite was quickly platted and saloons, an assay office, and a boardinghouse were built. A stage line to Ely was set up. The proximity of the Nevada Northern Railroad helped aid the camp in its growth. The boom was short-lived, and by 1909 the mill and mines had all stopped. Total production from 1902 to 1910 was $14,500. The Cuba Mine was reopened in 1911 and was the sole producer of the district until it closed in 1960. Total production for the Granite District was $201,000. Scattered ruins remain in the canyon to the west of present-day Campbell Ranch.

Bonita (Snake District)

DIRECTIONS: *Located 5 miles south of Lehman Caves. Site is virtually inaccessible except by foot.*

The initial discoveries at Bonita took place in 1869, but it wasn't until 1913 that measurable production took place. The Tilford Mine was discovered in spring 1913, and a two-ton experimental mill was built to treat the scheelite ore. The camp soon had a population of 25 and a pair of saloons. In 1918 the Uvada Tungsten Company, working the Pilot Knob claims, built a 20-stamp mill. The mill was doomed to failure, however. Not only was it poorly made but the mines couldn't supply enough ore for a 5-stamp mill, let alone a 20-stamp one. This financial disaster crushed Uvada Tungsten, and the company went into bankruptcy in 1919. Bonita was abandoned by 1920, after production of only $10,000, which did not even come close to covering the cost of the mill disaster. No activity has taken place here since, and only mill ruins mark the site.

Bothwick (Botha Creek)

DIRECTIONS: From Keystone Junction, head west on U.S. 50 for 15.6 miles. Exit right (this is the old Bothwick toll road) and follow for 3 miles. Bear right and follow for 1 mile to Bothwick.

Bothwick was a stage station on the Lane City–Hunter stage line from 1877 to the early 1880s. The stage ran via Hercules Gap. William Botha built the station and constructed the Bothwick Toll Road. Botha made quite a bit of money during the Hunter boom, but his success came to an abrupt end when he was ambushed and killed by Indians. Because of the excellent location, families moved to Botha Creek once the Indian troubles had ceased. When changes were made in Homestead Laws in 1909, thirteen families moved here. A school was built and homes constructed. Most of the families had left by 1940 and ten years later, Botha Creek was empty. One building, believed to be the schoolhouse, struggles to stand at the site. Only a faint stone foundation marks the site of Botha's Station. Other ruins of old homesteads also are scattered along the banks of Botha Creek.

Broadway

DIRECTIONS: Located 2 miles south of Eight-Mile Station. No roads go to this site.

Broadway was a short-lived camp that sprang up during spring 1893. Very little is known about the camp, although a post office did open there on July 19, 1893. Emily Bently served as postmistress until the office closed on September 7, 1894. No other substantial information is available except that a ranch was active in the area during this period. Whatever its history was, evasive Broadway quickly faded from memory, and nothing at all marks the site today.

Buck Station

DIRECTIONS: From Illipah, head west on U.S. 50 for 20.2 miles. Exit right onto the old Elko-Hamilton stage road (very poor, heavily rutted) and follow for 11 miles. Bear left at fork and continue for 9 miles to Buck Station.

Buck Station was an important stop on the Elko-Hamilton stage line. The station was the scene of bustling activity during the late 1860s as travelers rushed from northern Nevada to Hamilton. The Hill-Beachy freight line used Buck Station as a place to switch horses. There was a constant stream of adventure seekers, and hundreds of heavily loaded freight wagons rumbled

Old Buck Station has been virtually destroyed by the elements.

through the station. Buildings were constructed, including a large horse barn and a boardinghouse. Dining facilities were also available for tired travelers. All was not safe and secure at Buck Station, however. In May 1869, the Wells-Fargo stage was held up. $40,000 was taken, and four men were killed. The money was never recovered, and legend has it that the treasure is buried somewhere near the station. When the robbers were caught, only a few miles away, they had already hidden the money.

Once Hamilton began to decline in 1870, Buck Station lost its importance, and by the late 1870s the station was no longer used. In the 1880s, a small and very successful ranch began operations, continuing in business until the 1930s. Today there are extensive and fascinating remains at Buck Station. The partially standing ranch house gives an aura of ghostliness. The impressive horse barn is an excellent example of early frontier architecture. Other stone remnants abound at the site. One of the most interesting relics is a Model A Ford, slowly rusting away next to the old stage road. Buck Station, a must for any ghost town enthusiast, is well worth the tough trip.

Bull Spring Station (Bull Creek)

DIRECTIONS: From Eberhardt, continue south and follow this graded road for 1.5 miles to Bull Spring Station.

Bull Spring Station was an obscure stop on the Elko-Pioche stage run. The station was located on Bull Creek Ranch and used primarily as a horse-exchange place, although meals and lodging were available. Once the stage line was abandoned in the early 1880s, Bull Creek became solely a ranching operation and is still active today. Old buildings, including one believed to have been the station house, still stand on Bull Creek Ranch.

Butte Station (Bates) (Thieves' Delight) (Robbers' Roost)

DIRECTIONS: From Cherry Creek, head west on good road for 7.5 miles. Exit left and follow this road, which gradually bends south, for 7 miles. Exit right and follow for 6 miles to Butte Station.

Butte Station served as a station for the Pony Express, Overland Mail and Telegraph, and Chorpenning Mail Line. Chorpenning used the station during 1859, and the Pony Express took it over in spring 1860. The Pony Express station was called Bates, although most writers list it as Butte. Butte Station was built in 1862 when the Overland line was improved. Indian troubles plagued Butte Station. On June 27, 1860, the station was burned. The station was quickly rebuilt, and Sir Richard Burton gave this report when he visited the station on October 5:

> About 3 a.m. this enjoyment (riding on the stage) was brought to a close by arriving at the end of the stage, at Butte Station. The good station-master, Mr. Thomas, a Cambrian Mormon . . . bade us kindly welcome, built a roaring fire, added meat to our supper of coffee and doughboy, and cleared by a summary process amongst the snorers, places for us on the floor of Robber's Roost or Thieves' Delight, as the place is facetiously known throughout the countryside.
>
> It is about as civilized as the Galway shanty, or the normal dwelling place in Central Equatorial Africa. A cabin fronting east and west, long walls thirty feet, with portholes for windows, short ditto fifteen; material, sandstone and log ironstone slabs compacted with mud, the

whole roofed with split cedar trunks reposing on horizontals which rested on perpendiculars. Behind the house rested a corral of rails planted in the ground; the enclosed space a mass of earth, and a mere shed in one corner the only shelter. Outside the door—the hingeless and lockless backboard of a wagon bearing wounds of bullets—and staples, which also had formed parts of locomotives, a slab acting stepping-stone over a mass of soppy black soil strewed with ashes, gobs of meat offals, and other delicacies. On the right hand a load of wood; on the left a tank formed by damming a dirty pool which had flowed through a corral behind the Roost. There was a regular line of drip distilling from the caked and hollowed snow which toppled from the thick thatch above the cedar braces.

The inside reflected the outside. The length was divided by two perpendiculars, the southern most of which, assisted by a half-way canvas partition, cut the hut into unequal parts. Behind it were two bunks for four men; standing bedsteads of poles planted in the ground, as in Australia and Unuamwezi, and covered with piles of ragged blankets. Beneath the framework were heaps of rubbish, saddles, cloths, harness and straps, sacks of wheat, oats, meat and potatoes, defended from the ground by underlying logs, and dogs nestled where they found room. The floor, which also frequently represented bedsteads was rough, uneven earth, neither tamped nor swept, and the fine end of a spring ooz-

A partial wall remaining at Butte Overland Station.

> ing through the western wall kept part of it in a state of eternal mud. A redeeming point was the fire-place, which occupied half of the northern short wall . . . there was no sign of Bible, Shakespeare or Milton; a Holywell Street romance or two was the only attempt at literature. Instead, weapons of the flesh, rifles, guns, and pistols, lay and hung about the house, carelessly stowed as usual.

Don Salisbury, the tender at Egan Station, came to Bates Station and stayed for ten months to help rebuild it after the Indians burned it. He was born on October 25, 1841, in Illinois, to Catherine Smith Salisbury, sister of the Mormon prophet Joseph Smith.

The Overland Mail and Telegraph Company left the station in 1869. While some people continued to live nearby, the area was abandoned by the mid-1880s. Today parts of the Overland station are still standing. Sections of the walls and part of the huge fireplace are all that remain of the once-popular outlaw hangout. Bates, the earlier station, was located four miles north at Pony Spring. Foundations mark the site.

Cherry Creek

DIRECTIONS: From McGill, head north on U.S. 50 for 33.3 miles. Exit left on Nevada 489 and follow for 8.2 miles to Cherry Creek.

Cherry Creek has been one of White Pine County's best-producing districts for well over 100 years. It all began on September 21, 1872, when Peter Corning and John Carpenter from nearby Egan Canyon located the Tea Cup claims. Within one year the list of mines included the Star Pacific, Exchequer, Flagstaff, Corey, Eagle, Mary Anne, Black Metals, Mother Lode, and Bull Hill. The Cherry Creek boom was on. By spring 1873 the town had a population of 400. A 5-stamp, 25-ton mill, the Thompson, was constructed next to the Tea Cup Mine. Buildings, including a livery stable, a blacksmith shop, a $2,000 hotel, boardinghouses, restaurants, and—most important to many miners—more than twenty saloons, quickly sprang up in Cherry Creek as the boom gained momentum. Peter Newman constructed a brewery in Egan Canyon to supply the thirsty Cherry Creek saloons. Because of the booming Cherry Creek economy, Wells-Fargo opened a station in 1873. A post office also opened. Another small mill, the Flagstaff (Henry Lyons, superintendent), started on May 17. All this bustling mining activity began to fade in 1874. In the town's elections that year, more than 500 ballots were cast, but even then most of the mines and both mills were struggling. By 1875 most had closed and only limited production continued.

Interior of the Elk Saloon at Cherry Creek. (Nevada State Museum)

In 1880 Cherry Creek revived and began its biggest boom. Rich new finds were made in the Exchequer and Tea Cup mines. Soon after, additional veins were discovered in the Star Mine. By the end of 1881 the mines each employed close to 200 men. A 20-stamp, 100-ton amalgamation mill was moved from Hamilton (the Dayton Mill) to the Star Mine. The Star Mill was started in July 1882. Other mills put into operation included the 50-ton Exchequer and the 5-stamp Tea Cup.

Cherry Creek quickly became the largest voting precinct in White Pine County. The post office was reestablished and operated out of a mercantile store owned by D. H. Gray, a prominent White Pine politician, and Daniel Collins, who also acted as postmaster. At its peak in 1882 Cherry Creek had a transient population of 6,000 and about 1,800 permanent residents. The town now had an amazing 28 saloons, keeping the brewery running at ca-

A close-up of the old Cherry Creek store.

pacity. In addition, a wide variety of mercantile stores were operating, and a Woodruff and Ennor stage to Toano (Elko County) was set up. By popular demand, a track for horse racing was built three miles south of town. The racetrack, a source of great civic pride, boasted a huge grandstand, stables, and a mile-long track. Horses came from as far away as Missouri to race here, and the track had a reputation of being the fastest in Nevada.

Literary enlightenment came to the town when the *White Pine News* was moved from Hamilton on January 1, 1881. By 1882, J. B. Williamson, owner of the Exchequer Mine, had shipped more than $1 million in bullion. Then the financial crash of 1883 stopped Cherry Creek's boom in its tracks. That, combined with poor management, forced the Star Mine and Mill to close in 1884. Soon after, the Tea Cup and the Exchequer also closed. Cherry Creek began a rapid decline. Despite its dwindling population, the town tried to challenge Hamilton for the county seat—but to no avail. A further blow occurred on July 24, 1884, when hoisting works at the Star Mine burned. By November only one saloon, Coqners and Boss, was still serving the shrinking populace. On August 15, 1885, the *White Pine News* left Cherry Creek and moved on to the boomtown of Taylor. While the major mines had closed, smaller ones were still producing. Cherry Creek continued to decline, however. A fire in August 1888 destroyed a section of the business district, causing $20,000 in damage. By 1890 Cherry Creek had a population of only 350. Three stages (to Wells, Aurum, and Ely) still ran from Cherry Creek, but another fire, in

February 1901, further depressed the camp. The fire, started by a man named Abe Kooken as he tried to put gas into a hot lamp, destroyed a major section of downtown Cherry Creek. Yet another, smaller, fire occurred in 1904.

Then, beginning in 1905, Cherry Creek experienced a revival. The Tea Cup (renamed the Biscuit), Exchequer, and Star mines were reopened, and two new mines, the National and the New Century, were put into production. An additional boost to Cherry Creek occurred on July 17, 1906, when the Nevada Northern Railway arrived. The population grew to about 450 before the revival faded in 1910. During the next few years, only leaseholders were active in the district. The Cherry Creek Silver Dividend Mining Company worked the Mary Ann Mine from 1917 to 1923 and produced $35,000. From 1902 to 1922, $701,000 was produced from the Cherry Creek mines. Several small mining companies came into the district during the early 1920s. The Penn-Star Mining Company (J. M. Murdock, president) worked nine claims and built a 100-ton flotation mill that ran until August 1921. The Tea Cup Mining Company (George Granopolous, president) began work in the district during 1918 but didn't begin production until the summer of 1919. The company controlled 22 claims, the most prominent being the Tea Cup Mine. A 100-ton flotation and cyanide mill was built in 1919, and a 7,250-foot tramway was constructed to link the Tea Cup Mine with the mill. An assay office and a number of other buildings were constructed at the mill. In 1924, the United Imperial Mines Company was organized and obtained control of the Exchequer, Imperial, and Star mines. These three mines contained over 18,500 feet of workings. The company also reopened the old Star Mill and converted it to a cyanide plant.

In 1927 the Nevada Standard Mining Company (J. Henry Goodman, president) entered the Cherry Creek District and began to purchase most of the district's claims. The company obtained the holdings of six companies active around Cherry Creek: Nevada British, Nevada Star, Glasgow and Western Exploration, United Imperial, Cherry Silver Star, and Cherry Star. Nevada Standard controlled 41 claims and actively worked the Star, Gray Eagle, Imperial, and Exchequer mines. As many as 200 men were employed in the mines and mill during the 1920s and 1930s. The mines had workings of more than 40,000 feet and had produced more than $10 million. The company worked the mines off and on until 1940. When the company folded, it had produced close to $2 million.

Since that time, leaseholders have always been active in the district. Even today, mining activity lingers in the Cherry Creek area and accounts for an increase in the town's population. Total production for the district is somewhere between $15 million and $20 million (estimates vary dramatically). Cherry Creek is definitely one of the best ghost towns in Nevada. Many buildings remain, including the school, several old saloons, and a couple of false-fronts. Until recently, the old Nevada Northern Railway station was located

One of the many interesting gravesites at the Cherry Creek Cemetery.

just east of town. The depot, which retained a rare and complete water tower, was moved to the White Pine Museum in Ely. The water tower has now collapsed and remains at the depot site. Cherry Creek has a fascinating cemetery, which contains many old wooden markers. Unfortunately the cemetery has been vandalized, and many markers were broken and scattered. About 20 residents live in the town, and one of the saloons is in operation. A fire a few years ago destroyed a group of beautiful old falsefronts, located in what used to be downtown Cherry Creek. Plenty of mine and mill ruins are located further up the canyon. Cherry Creek is a must for the interested visitor; plan to take more than a day to enjoy and explore the town's many and varied points of interest.

Claytons

DIRECTIONS: From Tippett, head east on good road for 11 miles to Claytons.

Little is known about the short-lived mining camp of Claytons. George Clayton, the founder of the camp, discovered some gold ore here in early 1880. Clayton lived in Huntington Valley, Elko County. Because of its isolation, the camp never grew. All supplies had to be brought in by private wagons. By spring 1881 Claytons had a population of 29, most of whom were employed in two mines, the Antelope and the Rees. The already small veins disappeared altogether during the summer of 1881, and by that winter, Claytons was completely abandoned. Total production was only about $2,000, and no attempts were ever made to revive the mines. Only one frame building was built. Today small mine dumps are the only markers of the site.

Cleve Creek (Kolcheck District)

DIRECTIONS: *From Cleveland Ranch, head south for 3 miles. Exit right and follow for 7 miles. At fork, bear left and follow for 1.5 miles to Cleve Creek.*

Cleve Creek was the site of a small mining camp that sprang up in 1923 after the discovery of the Kolcheck Mine. A small 4-ton amalgamation and concentration mill was built, and a camp of 15 people formed. The mine operated on a small scale until 1926, when it closed after producing about $15,000. The mine was idle until 1951 when it was retimbered and production resumed. During the next five years, 235 tons of gold ore and 32 tons of tungsten ore were shipped. No production has taken place since the late 1950s. Today only one wooden building remains, along with some mine ruins. A point of interest along Cleve Creek are two sets of Indian pictographs, one group at the fork of Cleve Creek and North Fork of Cleve Creek, the other about a mile and a quarter farther up Cleve Creek. Camping areas are located all along Cleve Creek, but the road is quite rough and includes creek crossings, which, during spring runoff or after summer cloudbursts, make driving hazardous.

Cleveland Ranch (Cleveland)

DIRECTIONS: *From Ely, take U.S. 50 south for 29.4 miles. Exit left onto good road and follow for 16 miles to Cleveland Ranch.*

During the mid-1860s, Abner C. Cleveland and Daniel Murphy became partners in a cattle venture. Cleveland borrowed $50,000 from Murphy to enter the partnership, and the two stocked the new ranch with Hereford cattle. Within a few short years, the ranch was the most successful in Spring Valley. Cleveland married Kate Peters in 1868 and two years later was elected to the Nevada State Senate. Murphy died in 1882, and Cleveland took over complete control of the ranch. A post office opened on July 24, 1882, and mail was brought from Gold Hill, Utah, twice a week by Tom Mulliner.

Early photo of Cleveland. The Xs indicate Mr. and Mrs. A. C. Cleveland. (Nevada Historical Society)

The post office closed in 1905 but reopened from 1917 to 1924. Cleveland was an important man on the political scene in Nevada. In 1890 he ran for the United States Senate but lost the election to William Stewart. In 1902 he ran for governor but lost again, this time to Governor John Sparks. He died the following year.

After Cleveland's death, his widow took control of the ranch. William Neil McGill, for whom the smelting town McGill was named, made numerous attempts to buy the Cleveland empire, which led to a bitter feud. During this period, Cleveland was the primary stop on the Aurum-Osceola stage, which ran twice a week. In 1909 Mrs. Cleveland sold the ranch to Thomas Judd, a Mormon bishop from Lund, for $100,000. She believed Judd when he said he would not resell to McGill. But only months later Judd deeded the land to McGill. Mrs. Cleveland vowed that she would kill McGill for his underhanded deed, but she never carried out her threat. Cleveland Ranch is still in operation today, and many of the original buildings remain. A few cowboys who died while working at Cleveland are buried in the Osceola Cemetery, including Decknos Kathan, a Canadian who died on March 24, 1899. One of the interesting aspects of the Cleveland site are the huge trees, about fifty of them, that line the road to the old ranch.

Cocomongo (Watsonville)

DIRECTIONS: From Cherry Creek, head south toward Egan Canyon for 4 miles. At four-way intersection, continue straight for 2 miles. Look for streambed on right. Follow this streambed (there may be water in it during spring months) for about 1.5 miles to Cocomongo.

Will Watson, W. B. Lawler, and Ernest Baker made a gold strike in 1903. The ore assayed at close to $400 per ton, and a camp of 30 formed. Access to the camp was extremely difficult, and water was hauled in from Egan Canyon. The main producer was the original discovery, the Joanna Mine. A new mine, the Pick and Gad, was discovered by Charlie Wah in 1904. In 1905 the Hartford Nevada Mining Company gained control of the two mines. Within months, the company merged with the Gold Canyon Mining Company. A 40-ton mill was built in 1906. The mill had four 1,000-pound stamps and a 50-horsepower gas engine. A tramway was built from the Joanna Mine to the mill, which ran for two years before closing. Hartford Nevada folded in the summer of 1908, and by 1909 Cocomongo had been abandoned. The mill was taken down in 1917 and moved to Piermont. Today the ruins of three wooden cabins remain, but not much else. Only the fanatic ghost town hunter will make it to Cocomongo. Finding it is an extremely difficult hit-or-miss proposition. The faint wagon road to the site fades in and out and is very tough to follow. If you are successful in finding the site, you will receive the reward of knowing that you are one of only a handful who have visited Cocomongo!

Cold Creek (Cupperville) (Simonsen)

DIRECTIONS: From Eureka, take U.S. 50 east for 14.5 miles. Exit left and follow good road for 33 miles to Cold Creek.

Cold Creek is one of the many long-lived Newark Valley ranches. The ranch was established in the mid-1870s, and on April 7, 1879, a post office was bestowed upon it. Nicholas Simonsen served as postmaster, and Cold Creek became the mail distribution point for the valley's ranches. In the early 1900s, during the mining boom in nearby Joy, Cold Creek was known as Cupperville. A school was active at Cold Creek for many years. On March 28, 1913, the post office was renamed Simonsen. The office closed on May 15, 1936. Cold Creek continues to be an active ranch. Several of the original buildings remain.

The Cold Creek school and students, 1905. (Northeastern Nevada Museum)

Conner's Station (Connor's Station) (Rosebud)

DIRECTIONS: *From Ely, head south on U.S. 50 for 22 miles to Connors Pass, site of Conner's Station.*

Little is known about Conner's Station, located on the old Ely-Pioche stage route. A quaint frame station house, complete with stables, was built here in 1876 by the widow Conners, who had been the cook on the Cannahan Ranch. The station was located at the top of 7,723-foot Connor Pass. Horse teams were switched here after the long haul up the mountain. Mrs. Conners prepared home-cooked meals for the travelers, and lodging was available for $1.50 per night. Water for the station came from nearby Rosebud Spring, which sometimes ran dry during the hot summer months. When the Ely-Pioche stage stopped running, Conner's Station was no longer needed. Today only some scattered rubble marks the site.

Duck Creek (Kent) (Success) (Peacock)

DIRECTIONS: *From McGill, head north on U.S. 50 for 5.3 miles. Exit right and follow good road for 5 miles to Duck Creek.*

Duck Creek was the site of a sizeable ranching settlement formed in the late 1860s. Long before the first white settler arrived, the area was the site of a tragedy. On May 4, 1863, Colonel S. P. Smith of Fort Ruby led the

Company K Cavalry into Duck Creek and killed 24 of a group of 26 Goshute Indians. Some ore was discovered in 1869, and the Enterprise and McDougal mining districts were formed. Ranching, however, soon became the main industry of Duck Creek. By 1872 there were ten or more ranches in the little valley. A post office opened on June 10, with Samuel Cauldwell as postmaster. On August 16, 1873, tragedy struck again when Cauldwell's daughter drowned while playing in Duck Creek. The entire population of close to 60 people attended the funeral. Cauldwell left the following year, and the post office closed on March 17.

In October 1882 Duck Creek had 18 registered voters: Ira Abbott, Byron Bird, Cyrus Chase, Patrick Coffee, John Cowger, Thomas Freehill, William Gallagher, Andrew Guilford, James Guilford, Henry Jones, Sr., Henry Jones, Jr., J. A. P. Jones, Jeremiah Kent, Nathaniel Kinsley, Micheal McHugh, Edwin Marks, Emille Meyer, and Alexander Ryan. The Duck Creek settlement had a consistent population of 50 for a long time. Beginning in 1890 mail was brought here twice a week from Ely. Crane Gallagher and his wife ran the post office. Gallagher also built bridges across Duck Creek to make it easier for supply wagons to come and go. The post office was renamed Kent, after an early settler named Jerry Kent, on June 20, 1899. It remained open until January 15, 1907. After the turn of the century, a twice-a-week stage line to Ely was organized. Some mining activity took place here beginning in 1905. The Success Mine, discovered by D. C. McDonald of Ely, produced lead, silver, and gold. The Lead King Mining and Milling Company of Utah leased fifteen claims in the Duck Creek district in 1913, and high-grade galena ore was located in 1916. The first ten carloads produced $20,000. Between 1905 and 1921 mining activity in the Duck Creek area yielded $165,000. The population of the area has fallen off in recent years, but four ranches are still active. The state-run Duck Creek Fish Hatchery is located here.

Eberhardt

DIRECTIONS: *From Hamilton, head east, then south, on poor road for 5 miles to Eberhardt.*

Eberhardt was one of the more important satellite towns near Hamilton. The camp was located on the southeast side of famed Treasure Hill. Initial discoveries were made in December 1867 when T. E. Eberhardt located the Eberhardt Mine. The mine had a fabulously rich ore chamber, which the federal mining commissioner reported as having walls and ceiling of silver. Work was begun on the California, or Stanford, Mill in early 1869. In June construction was started on another 30-stamp mill, the International. During construction of the mills, Eberhardt's population grew to more than 200. Businesses opened, and mail was brought from Hamilton three times a

The International Mill, looking north. (Bancroft Library)

week by Tom Starr. As mill construction progressed, Eberhardt continued to boom. The California Mill, designed by W. H. Patton and built by the Stanford brothers, cost $75,000 and was started on October 20, 1869. The mill had 30 stamps, four retort and two melting furnaces, 16 amalgamating pans, eight settlers, four agitators, and a 50-ton kiln. The Stanford brothers used the mill to treat ore from the California, Aurora South, Jim Stewart, Evening Star, Mahogany, and Poorman mines.

The International Mill took longer to build and was more expensive. The mill cost $257,000, and a tramway, at that time the largest in the nation, cost another $137,000. After a long delay while the tramway was completed, the mill finally began operations in May 1871. To solve the problem of occasional water shortages, the owners of the mill, the British-controlled Eberhardt and Aurora Company, bought Applegarth Springs, just north of town, for $150,000. Pipelines were laid and the mill thereafter always had a full supply of water. The pipeline also supplied the populace of Eberhardt. A post office (Adam Johnston, postmaster) opened on June 19, 1871, and was in operation until July 11, 1893. The milling camp had a consistent population of 150. Saloons, mercantile stores, and other businesses thrived here. Disaster struck on August 31, 1872, when the International Mill was destroyed by a suspicious fire. Ore was shipped to the Big Smoky Mill in Hamilton while the

International was being rebuilt. Master millright Perlia Rowe and chief engineer J. H. Montgomery supervised the reconstruction. More than 120,000 bricks and 189,000 feet of lumber were used. The mill was started in November 1873 and had six batteries of 5 stamps each, 16 pans, eight settlers, four retort furnaces, and a 56-by-16-foot dry kiln.

The Eberhardt and Aurora Mining Company expanded its holdings in 1873 when the White Pine Water Works and the Aurora Consolidated Silver Mining Company were purchased. While the California Mill closed and was dismantled in 1876, the International continued to operate. Eberhardt's population stood at 170 in 1880, but soon after, the town began a rapid decline. Mining became sporadic, and in 1885 the company closed down the mill and sold its property to the Eberhardt and Monitor Mining Company. During the next ten years, four other companies leased the property, but production was quite small. By 1897 only four homes and two inhabitants were left in Eberhardt. William Miles Read, the last leaseholder, worked the mines until 1904. Today the massive foundations of the California and International mills dominate Eberhardt. Building ruins are scattered throughout the site. The old tramway is visible in some spots on the side of Treasure Hill. The short-lived camp of Sunnyside was located near the large mine dump visible from Eberhardt, on the southeast side of Treasure Hill just above Eberhardt's ruins. Eberhardt is worth the trip because of the extensive remains of mills, which clearly show how these huge mills operated. The site presents unlimited exploration possibilities for photographers and ghost town enthusiasts.

Egan Canyon (Gold Canyon)

DIRECTIONS: From Cherry Creek, head south on poor road for 5 miles to Egan. The Pony Express site is located at the west end of the canyon, in a small open area.

Egan Canyon was the site of a Pony Express station and later served as a stop on the Overland Stage. As early as 1859, the canyon had been used for the Chorpenning mail service. The station was named after Howard Egan, overseer of the Pony Express line in this area. Egan Canyon was constantly plagued by Indian attacks, the most serious of which was the so-called Battle of Egan Canyon.

> On the 16th of July, 1860, the only men at Egan Canyon station were Mike Holten, stationkeeper, and Wilson, rider, who took the Express from Will Dennis who had my ride from Ruby east, and carried it to Schell Creek. The soldiers had left and the other three employees of the Express Company who had been there for a month past, were sent to work on other portions of the route, as we all supposed the Indian war

was over. But on the day referred to, about 80 of the renegade [Indians], who had fought under Leatherhead, in all their war paint, rode through Egan Canyon up to the station and demanded of the boys, flour, bacon and sugar. The boys handed out the provisions knowing it would not do to refuse. Mike then started out to gather the Express horses up and put them in the stockade corral, but one big Indian, who could talk some English, told Mike to go in the house, that the Indians would take care of the horses and them too after they had their feast.

Holten and Wilson were brave men, well armed and expecting to be massacred by the [Indians] after powwow was over, closed up their door and barricaded the only door and window they had in the log cabin with grain sacks, leaving a few chink holes to shoot through, determined to sell their lives dearly as possible. It was a trying time for those two men, but they had nerves of steel and expected to make several reds bite the dust before they lost their hair. They knew that it would soon be time for Dennis, the Pony rider from the west, to arrive and they thought as he did not show up that the Indians must have waylaid him and killed him, but such was not the case.

After Dennis came through Nipcut Canyon, which was steep and rocky, he rode fast with the Express until he came even with the knoll (behind the station), when he pulled up his horse for a moment to get his wind, as we usually would let our horse walk until we came in sight of the station. Dennis caught sight of the Indians before they saw him. He comprehended the situation instantly and whirled his horse out of sight of the redskins. He passed the soldiers who were on the road to Camp Floyd, about 5 miles back, so he rode back as fast as possible to the command and informed Lt. Weed of the situation, who immediately started for Egan Canyon with 60 dragoons. They rode fast until they got to the knoll.

Orders were then given to Corporal Mitchell to take 20 men and go to the mouth of Egan Canyon and cut off the retreat of the [Indians], but in the excitement of the moment Mitchell got his orders mixed up and instead of going to the mouth of the canyon, he led his men around the east side of the knoll and charged the Indians. As soon as Lt. Weed heard the shooting he rode around the west side of the knoll and charged right into the fight.

When Holten and Wilson saw they were about to be rescued they did rapid shooting themselves. The fight was soon over; 18 Indians fell to rise no more, and the rest of the [Indians] made their escape through the canyon. Had Corporal Mitchell not made any blunder the whole band of [Indians] would have been killed. The soldiers got 60 of the Indians' horses; three soldiers were killed and several wounded, Corporal Mitchell receiving 3 shots, one through the back. He recovered from his wounds, but died about 6 months afterwards.

After that battle the Indians sued for peace, but did not keep it, as they committed many murders on the road after that and during the next summer.

The Indians returned in October, bent on revenge, and burned the station, killing the station tender. After the Pony Express folded, the Overland Stage incorporated Egan Canyon into its route. The first station tender was Dan Salisbury, who was paid $50 per month in gold for his services. The cabin he worked in was furnished with a stove, cooking utensils, prepared flour, and bacon, all brought by the stage company. Salisbury employed two assistants, a hostler to care for the horses and change the teams and an Indian to grease and clean the harnesses. Indians, now peaceful, constantly visited the station, looking for free handouts.

Gold was discovered by a Captain Tober and Company C of Fort Ruby in September 1863. A mining district was organized on September 23, and the camp of Egan Canyon began to boom. Among the early producers were the Gilligan (discovered by John O'Dougherty on February 25, 1864), Hope (October 3, 1863), Gold Canyon (September 29, 1863), and Jenny Lind (initial assays of $3,500 per ton) mines. The Social Mining Company and the Steptoe Mining Company became active in Egan Canyon during late 1863 and were the two main producers in the district. Social Mining built a 5-stamp mill, which began production in October 1864. The mill had five 650-pound stamps, three Varney pans, two settlers, and a capacity of 5 tons a day. Before the mill closed in 1868 it produced $80,000. By fall 1865 two other mills were in operation: the 5-stamp Roose and the 10-stamp Stephen, built by Steptoe. By 1866 Egan had become a small town, with stores, a school, residences, and a post office that opened on April 13. During 1866 the Stephen Mill was expanded to 20 stamps and renamed the Social and Steptoe. The two mining companies merged, becoming the Social and Steptoe Mining Company. The company employed more than 100 men to work the mines and mill, cut wood for the mill, and make equipment in the company's three blacksmith shops. The mill closed in May 1870, and the mines followed suit in July.

Egan Canyon was revived in 1872 when a General Rosecrans reopened several mines. Rosecrans and Bart O'Connor organized the San Jose Mining Company, and construction was begun on a new 20-stamp mill. The Egan Mill was the first mill in Nevada to be run entirely by waterpower. Before the mill was completed, ore from the mines was shipped via the Woodruff and Ennor stage line to Wells. Postmaster Patrick Logan owned most of Egan Canyon's businesses, including the camp's only store, saloon, and boardinghouse. He also owned Egan Canyon's water rights and townsite. Two other small mills joined the Egan Mill during 1873. The 5-stamp Wide West Mill and the 80-ton Exchequer Mill were built. The Wide West was owned by San Jose Mining. At the mill stood an assay office, a two-story mansion that served as the company president's home, and a telegraph office. Rosecrans spent more than $250,000 developing his Egan Canyon holdings during the 1870s. The most important mines in the district included the Gilligan, Jenny Lind, Wide West, San Jose, Centennial, Pine Tree, and Eastern mines. Egan experienced a mining slowdown in the late 1870s, and the post office closed on March 19, 1878. Despite the slowdown, the work force consistently exceeded 100 men. In August 1881, 110 were employed, 50 at the Egan Mill. The main vein pinched out in late 1882, and Egan fell on tough times. The San Jose Mining Company folded in 1883 after producing close to $400,000. All mining activity ended by 1885, and only leaseholders worked the districts during the next ten years.

In 1896 the newly organized North Mountain Mining Company purchased the old San Jose holdings. The focus of the company's exploration was the Gilligan claim group. The company enjoyed modest production through the 1920s. In the mid-1920s, the Goshute Mining Company leased the property, but after five years of low production, Goshute left the district. Some additional leasing took place in the late 1930s, and a little over $100,000 was produced. Egan Canyon has been quiet ever since. Remains of the settlement are scattered throughout the canyon. The Pony Express station site and soldier cemetery are located at the western mouth of Egan Canyon. Mill ruins and the Egan townsite are at the eastern mouth. At the station site, faint foundations remain but are covered by thick sagebrush and so are difficult to find. The soldier cemetery is located to the north of the road. Collapsed wooden cabins and mine ruins are scattered along the canyon road. Not much remains of the Egan camp and mills. Some mill foundations are still evident, but the townsite has been obliterated by later leasing activity. The road to both Egan Canyon sites is quite rough, so beware.

Eight-Mile Station (Prairie Gate) (Spring Station)

DIRECTIONS: *From Tippett, take Nevada 2 east for 20 miles to the Georgetta Ranch, site of Eight-Mile Station.*

Eight-Mile Station was a stop on the Overland Stage, named for its distance from Deep Creek, Utah. A simple mud-and-log-cabin station house was built here in July 1861. The Pony Express used the station during its last months of existence. It was here that the Goshute, or Overland, War began when Chief White Horse attacked the Overland Mail coach on March 22, 1863, near the station and later burned Eight-Mile Station. The attack proved to be the catalyst for a war that stretched 225 miles along the Overland Route, from Schell Creek to Salt Lake City. The war cost the Overland Stage Company sixteen men, seven stations, and 150 horses. The station was rebuilt in

1864 and served as a horse-exchange stop until 1869. No activity took place until the turn of the century, when Clel Georgetta, a prominent Nevada sheepman, settled here and established Eight-Mile Ranch. In 1938 Georgetta sold his ranch to the government for the Goshute Indian Reservation. Today the location of the old stage station is disputed. Many in the area believe that an old collapsed log dugout with a stone corral is the original station.

Ely (Murray Creek Station)

DIRECTIONS: Located on U.S. 50.

While discoveries were made as early as 1864, the town of Ely didn't form until 1870. Most mining was based in Mineral City, a few miles to the west. The original water rights on Murray Creek were controlled by J. R. Withington, but when George Lamb platted the Ely townsite in 1870, Withington relinquished the rights. The townsite, with only a few cabins, was more of a stage stop than a town. The first real house in town was built by Harry Featherstone, who ran the Murray Creek Station and post office. Featherstone also constructed a restaurant and a small hotel. Later he sold all of his holdings to R. A. Reipe, who enlarged the hotel and named it the Ely Hotel. The town was first called Ely during 1878, in honor of Smith Ely, presi-

Ely was still in its infancy in 1888. (Special Collections, Library, University of Nevada, Reno)

dent of the Selby Copper Mining and Smelting Company. The company built several smelting furnaces in Ely and gave the town its first real start. A post office opened in November, but Ely progressed very slowly. The population hovered between 25 and 30 until 1885.

In 1885, $5,000 was appropriated for construction of a courthouse. While a number of new stone buildings were built, most of the pre-1885 buildings in Ely were made from lumber salvaged from Ward and Taylor, nearby towns that had fallen on hard times. Sol Hilp, a prominent Nevada businessman, opened a store here. Then he ventured into politics, and his vast fortune soon vanished. Telephone service, along with light and power, arrived in 1886, controlled by A. P. Crompton. After the state legislature designated Ely as the new White Pine County seat in 1887, the town grew steadily until, by the end of the year, it was a bustling business center. Wells-Fargo opened an office to manage the flow of bullion through the district. The *White Pine News* moved to Ely in October 1888, from the fading town of Taylor. The paper, managed by W. L. Davis, remained in Ely until September 1907.

By 1890 Ely had a population of about 300 and was serviced by four stage lines: Ely-Eureka (six weekly trips), Ely–Frisco, Utah (two trips), Ely-Sunnyside (one trip), and Ely–Cherry Creek (two trips). In addition, a mail stage came to Ely seventeen times a week. Ely's economic prosperity was directly tied to the success of the mines located near Mineral City, Ruth, Kimberly, and Veteran. While Ely continued to grow, it wasn't until 1906 that the town really began to boom. Glowing reports of the richness of the Ely district attracted hundreds of hopeful people, including famous Nevadan Tex Rickard, who spent $15,000 to build the Northern Hotel.

The event of 1906 was the arrival of the Nevada Northern Railway on September 29. A big railroad-days celebration was planned, but the train arrived a day early! Mark Requa drove a special copper spike to complete Ely's economic lifeline. By the end of 1906, close to 5,000 residents were packed into Ely. The population explosion necessitated the building of a $35,000 schoolhouse in 1907. The financial panic of 1907 threatened to slow Ely's growth, but completion of the McGill Smelter in 1908 brought the town to new heights. Then low copper prices after World War I did slow growth. The copper mines closed, and the town's population shrank significantly. The mines were reopened during the 1940s and Ely revived. Now that the mines and McGill Smelter have once again closed, Ely is experiencing tough times once more. The town still has about 4,500 residents, and with new developments at Ward, Taylor, and Hamilton, and a new prison, Ely is looking optimistically toward the future. Hope springs eternal among the citizens that the copper mines will reopen and Ely will boom once again. Total production for the Ely district copper mines stands at more than $1 billion, making it the most productive copper district in Nevada history. When visiting Ely, be sure

Railroad Day in Ely, September 1906. (Nevada Historical Society)

to visit the White Pine Museum and the East Ely Railroad Depot Museum. The latter is located at the old East Ely depot, where train rides are offered. These two facilities offer an excellent look at White Pine County's glorious past.

Fort Ruby (Camp Ruby) (Hobson)

DIRECTIONS: *From Illipah, take U.S. 50 east for 1.5 miles. Exit left and follow this good road for 52 miles to Fort Ruby Ranch.*

"Uncle Billy" Rogers was the first settler in beautiful Ruby Valley. He arrived in 1859 and built a trading post. Rogers had a large garden and readily supplied visitors with a hearty meal. Rogers's post was incorporated into the Pony Express line in 1860. After the Pony Express folded, Overland Stage took over the station and operated it until 1869. Because of high prices for supplies around Camp Ruby, the Overland Company organized a huge farming district and supplied foodstuffs for most of Nevada's overland stations. Indians plagued the stage line during the 1860s, and it was decided that a fort was needed to protect the stage line and nearby settlers. Fort Ruby was established here on September 4, 1862, and a six-mile-square reservation was organized. The fort commander was Colonel Patrick Edward Connor. Fort Ruby was manned by Companies C and F, Third Infantry, of the California Volunteers, with Major Patrick A. Gallagher in charge of the troops. During the first few months, the men were kept busy erecting the company's winter quarters and the commander's huge stone house. The fort had a band, but to save expenses it was mustered out on September 23, to the disappointment of the troops. On October 2, Colonel Connor and the main force of the Third Infantry left Fort Ruby for Fort Douglas (near Salt Lake City). The squads led by Captains Izatus Potts and John H. May were left to man the fort, and Major Gallagher was appointed company commander. In November a group of unidentified culprits sneaked into the fort and stole ten of the troop's finest horses. Gallagher, Potts, and forty men gave chase, but the horses were never recovered.

Conditions at Fort Ruby were less than ideal. The fort was labeled as a "fever breeder and a natural hospital filler." Three soldiers died within the first couple of months. The attack on Eight-Mile Station on March 22, 1863, signaled the beginning of the Goshute War. The soldiers were constantly on missions, trying to protect White Pine County's residents. Captain S. P. Smith and Company K were assigned duty guarding the mail routes, while Companies E and B maintained patrols elsewhere. On May 1, Captain Smith and 87 men left the fort and began active campaigns to end the Indian threat. On May 4, 29 Indians were killed; on May 6, 23 more; and on June 20, another 10.

Fort Ruby, complete with company cannon. (U.S. National Archives)

In July, Major Gallagher, who had been wounded at the Battle of Bear River, was relieved of duty. He was replaced by Lieutenant Colonel J. B. Moore.

The Western Bands of Shoshone Indians signed a treaty at Fort Ruby on October 1, 1863, which ended the hostilities in eastern Nevada. The treaty resolved travel routes and Overland Stage lines through Shoshone land. The Indians agreed to let white settlers explore the land for gold and silver, but they received little in return. In the fall of 1864, Moore and his men were replaced by Captain G. A. Thurston and Company B of the First Infantry of Nevada Volunteers. They, in turn, were replaced by John Franklin Trout and the Ninth U.S. Infantry. With the end of Indian troubles came the end of the need for Fort Ruby. Most of the garrison was moved on March 10, 1869, to Camp Halleck, in Elko County. The property was sold to Thomas Short of Ruby Valley at a public auction on June 30, and the fort was officially abandoned on September 20. Some buildings were moved to the Cave Creek Ranch (Elko County), and graves in the cemetery were moved to Carson City. The Overland Stage left in 1869, and the huge farms were divided and sold at public auction shortly after the Fort Ruby abandonment.

A ranch began operations at the old fort site in early 1870. In 1898 the small town of Hobson, named for Admiral Richmond P. Hobson of Spanish-American War fame, was established at Fort Ruby Ranch. A post office, with Joseph Tognirri as postmaster, opened here on August 7, 1902, and served the many ranches in southern Ruby Valley until November 30, 1936. While a few buildings on the ranch date from the 1860s, no buildings from Fort Ruby still exist. In July 1992 the last two remaining buildings, the officers' quarters and a barracks, burned. The nearest firefighting equipment was far away, and the buildings were totally destroyed before firefighters arrived. Another valuable page of Nevada history had been lost. Now only stone foundations mark the Fort Ruby site. The old Pony Express and Overland Stage stations are located a short distance from Fort Ruby, designated by a historical marker and stone ruins. The old log station house was moved to Elko in 1960, where it is on display at the Northeastern Nevada Museum.

Glencoe (Well Annie)

DIRECTIONS: From Tippett, take good road directly east and follow for 23 miles. Exit right and follow for 1.5 miles to Glencoe.

Initial discoveries at Glencoe took place in 1867 when a group of Overland Stage employees located silver claims here. It wasn't until 1868 that development began and the camp of Glencoe came into existence. The Mammoth Mine was discovered in June 1868. In spring 1869 a large group of men from Kern County, California, moved to Glencoe. By summer of that year the

Pinyon, Buttercup, Sentinel, Spartan, Artic, Goshute, and East End mines had begun production. Glencoe began to attract attention, and the Nevada state mineralogist came in 1869 to survey Glencoe's mines.

One of the more prominent Glencoe residents was John Tippett, founder of nearby Tippett. Tippett and Frank Bassett were the discoverers of the Glencoe Mine. The two were once offered $100,000 for the mine, but refused to sell. Unfortunately the mine never came close to producing that much ore. Tippett and Arta Young ran the one store at Glencoe. While mining activity declined during the 1870s and early 1880s (only 34 residents were here in 1881), Tippett remained, working his claims. In 1890 Tippett and Young discovered the Well Annie Mine. The following May, Tippett was made the recorder of the Glencoe Mining District. Forty new claims were made and two new mines, the Harrison and the Paymaster, began production. A post office, with Tippett as postmaster, opened at Glencoe on September 2. In December the Well Annie Mining and Milling Company was organized. Oscar Hardy was appointed president, and 250,000 $1 shares were printed. Tippett and Young were each given 87,500 shares for rights to the mine. In May 1892 an eighteen-inch-wide vein was discovered and showed values of 40 to 600 ounces of silver per ton. Close to 50 men were employed by the company, and the camp of Glencoe reached its peak population of 75. By summer 1894 most mines had been worked out, and by fall, mining activity had ceased. The post office closed on October 17, and Glencoe was abandoned by winter. Today only faint ruins and tailings mark the site.

Hamilton (Cave City)

DIRECTIONS: *From Ely, take U.S. 50 west for 37 miles. Exit left and follow for 9 miles to Hamilton.*

While discoveries were made on Treasure Hill in late 1867, the terrain left much to be desired for a proper townsite. The search for a better site led to the organization of Cave City in May 1868, at the base of Treasure Hill. The new town inherited its name from the nearby hills, which were honeycombed with caves. The name was soon changed to Hamilton, after W. H. Hamilton, who, along with Henry Kelly and Ed Gobin, laid out the townsite. By June, the town had a population of 30 and one business establishment, a saloon. Then, rich discoveries on Treasure Hill created "White Pine Fever," and during the ensuing months, a huge rush to the district took place. A post office opened on August 10, and by winter Hamilton's population had swelled to 600. Businesses included a lodging house, a restaurant, and four saloons.

Once the spring thaw began, Hamilton's boom was on. In spring 1869 the town had a floating population of more than 10,000. Stage lines were running

The old Wells-Fargo office at Hamilton.

to Hamilton on a regular basis by summer, the four major ones being Wells-Fargo (Elko-Hamilton twice a day, beginning operations in February 1869), Wilson Brothers (triweekly stage from Austin, beginning April 1, 1869), Len Wines (stages from Austin and Elko to Hamilton and Treasure City, December 1868), and Pacific Union Express Company (Elko-Hamilton, February 1869). The last two merged in spring 1869 and added stages to keep up with the demand for goods.

When White Pine County was organized in March, flourishing Hamilton was selected as the county seat. A $55,000 brick courthouse was completed in the fall, and by the next summer Hamilton boasted a population of 12,000 and was incorporated. Main Street became lined with one- and two-story stone buildings. At the peak of White Pine Fever, there were close to *100* saloons (which meant the construction of a couple of local breweries), 60 general stores, and many other businesses. The town also had theaters, dance halls, skating rinks, a Miners' Union Hall, and fraternal orders.

Literary enlightenment came to Hamilton on March 27 when the *Daily Inland Empire* began publication. James Ayers and Charles Putnam ran the paper, which had a fierce rivalry with the *White Pine Daily News*. The *Daily Empire* lost the battle on April 10, 1870, although it was revived in October and November for the governor's race.

Because of the huge population crush, and in anticipation of an increase to as many as 50,000, a group of San Francisco investors organized the White

Pine Water Company to bring a water supply to Hamilton. The project employed 150 men to lay a twelve-inch pipe along a three-mile route to Illipah Springs and cost $380,000. The first water flowed through the pipeline on August 14, 1869. When the population decreased, the need for the 2.5 million gallons per day provided by the pipeline disappeared. The White Pine Water Company soon folded, and Stanford Mill interests acquired its property at a sheriff's auction on April 14, 1870. While Hamilton was not the center of mining activity, several mills were built near the town. Rothschild's Smelting Works was built in 1869 at a cost of $60,000. Its three furnaces, housed in a 209-foot-by-30-foot brick building, had a 20-ton capacity and 45-foot smokestacks. Other mills in the area included the 10-stamp Nevada (also known as the Dunn and McCoore, started May 20, 1869), the 20-stamp Big Smoky (also known as the Treasure Hill, built at a cost of $60,000 and started May 24, 1869), and the 24-stamp Manhattan (started June 1, 1869).

During the peak of the White Pine rush, close to 200 mining companies were active in the district. Most were just speculators, and once the uncertainty about the mines' potential became clear many of them left the district. Hamilton then experienced a depression and residents left as quickly as they had come. The 1870 census showed that population of the once-bustling town had shrunk to 3,915.

Hamilton, during its declining years. (Special Collections, Library, University of Nevada, Reno)

Remains of the Withington Hotel in Hamilton, shortly before a strong earthquake collapsed its walls. (Northeastern Nevada Museum)

Hamilton's decline was hastened by a disastrous fire on June 27, 1873. An owner of a failing store, discouraged by prospects for the future, set fire to his establishment to collect insurance. The fire spread throughout the business district and caused $600,000 in damage. Hamilton's slide continued, and by late 1873 the population had shrunk to 500. By 1880 only three saloons and two stores were still in operation. Another fire struck the town on January 5, 1885, destroying the courthouse and much of what remained in the dying town. And in 1887 Hamilton was dealt its worst blow when the county seat was moved to Ely. Hamilton was not quite dead, but only a few people inhabited the scattered buildings that remained. There was leasing activity on Treasure Hill, which kept a few small businesses running, but not much else. The population wasn't large enough to support the post office, and it closed on March 14, 1931, pounding the final nail into Hamilton's coffin.

Until recently, several stone buildings, most notably the two-story Withington Hotel, were still partially standing. But an earthquake, and humans, finally crumbled those buildings. Only rubble and pieces of stone walls now remain to mark the once-mighty Main Street. A number of smaller wooden buildings that somehow escaped the fires are scattered throughout the site. A trip to the Hamilton Cemetery, just north of town, is a fascinating exposure to the history of the area. The town revived in the early 1980s when extensive mining operations began on Treasure Hill. Once again Hamilton is the base for mining companies and has about 75 residents. Although heavy equipment abounds in Hamilton, care has been taken not to disturb the ruins. However, huge leaching pads and aluminum buildings tend to detract from the historical flavor of the town. Nevertheless, Hamilton is worth the trip. Many other ghost towns from the age of White Pine Fever are located around Treasure Hill. Plan at least a day to enjoy the beautiful White Pine District.

Hogum

DIRECTIONS: From Osceola, head west for 3 miles. At U.S. 50, take unpaved road heading south and follow for 1 mile. Exit left and follow poor road for 2 miles to Hogum.

Initial discoveries at Hogum were made in 1879 by T. Boone Tilford. He staked the Gaby-Buntin claim, but the camp of Hogum didn't form until the late 1880s. Three mines, the Mayday (130-foot tunnel), the Drummer (18-foot shaft), and the Serpent (50-foot and 80-foot tunnels) were developed and produced ore values as high as $400 per ton. By 1891 a small camp of 25 surrounded the Hogum mines. Mining ceased around the turn of the century, and it wasn't until 1911 that Hogum revived. The Hub Mining Company, headed by E. L. Fletcher, gained control of the district's claims and built a 30-ton mill and an open flume on Willard Creek. The company built a boardinghouse, a saloon, and a store to serve the 50 residents of Hogum. The promise of Hogum fizzled in the early teens and the camp was abandoned. A small revival took place in 1934 and 1935 when 25 men were employed to work the Hogum placers. Since that time, only occasional leaseholders have worked the district. Not much remains at Hogum. Until recently, Tilford's log cabin still stood, but now only rotting logs mark his homesite. Other ruins are scattered at the site, but nothing substantial remains.

Hunter (Carbonate)

DIRECTIONS: From Illipah, take U.S. 50 east for 11.5 miles. Exit left and follow for 3 miles. Exit right and follow for 4.5 miles, then bear left and continue on for 7.5 miles to Hunter.

Initial discoveries at Hunter were made by William Armstrong and Chris Decker on December 3, 1871. The early mines included the Horton, Tiger, Nebraska, and Arizona. These four mines produced $208,000 in their first year, but the Hunter District had to pay a high bullion tax. Hunter was the only district forced to pay this tax, which took $96,000 out of production figures. The Hunter mines were bought by George Kennedy of Cherry Creek, who built a smelter just north of the growing camp. During summer 1872 new mines included the Petersburgh ($1,100 per ton), Home Ticket ($47 per ton), Monte Negro, or Black Mountain ($300 to $2,000 per ton), Nebraska Tunnel ($850 per ton), and Emma ($350 per ton). The best mine was the Petersburgh, discovered by Nick Mars. The rich ore coming out of the district focused quite a bit of attention on Hunter, and the camp soon had a population of 50, plus a saloon, a blacksmith shop, a boardinghouse, and a general store. Hunter reached its peak in 1877 with close to 80 residents. Hunter had more than 40 houses, two stores, three restaurants, six saloons, two lodging houses, and a post office, with James M. Pay serving as postmaster.

In June the Crown Point Mining Company gained control of the district's mines and water and timber rights. Thirty houses were moved to Hunter from Cherry Creek and Schellbourne. The ore slowly ran out, and the Hunter furnace closed on July 12, 1884. At the time, the furnace employed 40 people. After it closed, most of Hunter's residents left. The handful left in Hunter in May 1886 were witnesses to murder when John Howlett, a prominent Carson City resident, was killed by Ed Crutchley. Crutchley was found guilty and hanged at Hamilton on December 31. He was the first man legally hanged in White Pine County.

Hunter continued to fade during the late nineteenth century. By 1900, the smelter had been wrecked, only one boardinghouse was still standing, and foundations were scattered throughout the site. P. A. Darraher started a small revival in 1904, and a two-story bunkhouse, a boardinghouse, and an office building were built. By 1905 the camp was full of tar paper shacks and 100 men were employed. The Hunter Mine, the district's only producer, was purchased by the Vulcan Mining, Smelting, and Refining Company in 1907. Between 1907 and 1916, when the company folded, a little more than $80,000 was produced. Leaseholders worked the district from 1920 to 1923, and the last activity occurred in 1948. Today only smelter ruins and decaying wooden shacks mark the site.

Illipah (Moorman Ranch) (Dutch Jake Ranch)

DIRECTIONS: *From Ely, take U.S. 50 west for 31 miles to Illipah.*

Illipah is the present name of a ranching complex that has had a rich history for well over 100 years. The ranch was originally established by Jake Medzgar, also known as Dutch Jake, soon after the Hamilton boom began. Captain William C. Moorman, a Confederate veteran, purchased the ranch from Medzgar and set up a toll road and tollgate along the "Narrows" heading from White River to Hamilton. Moorman, his wife, Pearl, and their five children made Moorman Ranch a prosperous venture. A post office opened here on March 22, 1898, with Pearl Moorman serving as postmistress.

Although the post office closed on November 30, 1913, Moorman Ranch has continued to be an active ranch. Buildings from the 1800s remain and are in use at Illipah.

Jacob's Well (Ivirs City)

DIRECTIONS: *From Cold Creek, continue north for 1.5 miles. Bear left and follow for 6.5 miles. Exit left and follow for 1.5 miles to Jacob's Well.*

Jacob's Well was a small way station for the Pony Express, although it was not originally on the route. Most likely the station served the line for only a few months. The station, named for General Frederick Jacobs, was a small stone cabin. A well was dug nearby to provide for humans and animals. During its existence, the station also served the Hill Beachy stage line to Hamilton and other White Pine camps. In 1863 the station was enlarged and meals and lodging were offered to travelers. The charge for a meal and a night's sleep was $1. In addition, a telegraph office was opened. The station continued to be active through 1869, serving the Overland Stage route. Some minor ore deposits were discovered near here, and a group of local prospectors organized the Huntington Creek Mining District, but nothing substantial was ever produced and the district had no recorded production. Today there is virtually nothing to mark the Jacob's Well site. Only a careful search will reveal some scattered stones remaining from the station. The well, now collapsed, is located nearby.

Joy (Bald Mountain) (Water Canyon)

DIRECTIONS: *From Cold Creek, head north for 2 miles. Exit right and follow for 3 miles. Exit left and follow poor road for 1.5 miles. Exit left again and follow for another 1.5 miles. Exit right for 4.5 miles. Ruins are scattered around this area.*

Silver ore was discovered in this remote area on August 13, 1869, by G. H. Foreman. A week later, on August 20, the Bald Mountain Mining District was organized. The richest area was the Free Metal Belt claims. By May 1870 the district had begun to make some noise. The Nevada Mine was the big producer, with ore assaying at $128 per ton. Other mines included the Genie ($40 per ton), Bismark ($80 per ton), Eastern, Silver Fringe, Alpha, Oneida, Mountain Queen, Winona, and Summit. It wasn't until 1875 that the mining camp of Joy was established. During the spring, Dr. J. M. Bailey gained a controlling interest in the district's mines. A camp of 25 formed, and freight lines from Elko and Halleck were set up. Joy's elevation of 7,400 feet and its location made access to the camp very difficult. Nevada mining magnate George Wingfield visited Joy but decided against investing in the camp. Discoveries in 1875 and 1876 included the Crown Point, Mountain View, Oddie Tunnel (one of 20 claims that made up the Blue Bell group), Redbird (J. G. Merritt), Carbonate (August Munter and Jacob Mayer), and Copper Basin group (25 claims owned by Simonson and Hannon of Skelton). A couple of placer deposits were discovered about half a mile west of Joy. Although they were promising, the lack of water proved to be a permanent hindrance.

By 1877 mining activity hit a slump, and from 1878 to 1897 only the Nevada Mine produced measurable amounts of ore. A small revival took place beginning in the spring of 1897, when a few old mines were reopened. A post office opened on May 6, with Jim Fulton as postmaster. The revival lasted only about a year and a half before investors withdrew support. The Joy post office closed on June 21, 1899, and the camp was pretty much abandoned until 1905 when the district's final revival began. Charlie Skeggs controlled 22 claims that made up the Bald Mountain Copper Basin group. During the summer of 1905 he sold his holdings to the newly formed Copper Basin Mining and Smelting Company for $300,000. By 1906 the producing mines in the district included the Dees Antimony, Dees Tungsten, Gold King, Oddie Tunnel, Copper Basin, Water Canyon, Carbonate, and Crown Point. The post office reopened on March 22, and D. O. Dees, district recorder, served as postmaster. Joy became a company town, owned and run by Copper Basin Mining and Smelting. Other companies were interested in the district, but Copper Basin already controlled all the claims. This closing of the district to other companies greatly limited Joy's growth. Copper Basin employed about 50 men, and Joy's population remained fairly constant at 75. Mines around Joy were consistent but not rich producers, and the cost of transporting ore across the great distances to mills and smelters kept profits from the Joy district quite low. Copper Basin folded in the summer of 1918, and with it went all the mines and all the stores, saloons, and other businesses. The post office closed on October 15, and by winter the camp was abandoned. A small revival took place from 1939 to 1942 when the Pioneer Mine was worked, and some leaseholder activity went on in the mid-1950s, but no production was recorded. Recently new activity has begun and plans include a major mining buildup. Today only scattered rubble, mine dumps, and hoisting works mark the site. The roads to Joy are very rough, and the trip should be attempted only with a four-wheel-drive vehicle.

Kimberly (Pilot Knob) (West Camp)

DIRECTIONS: *Located 1 mile from Ruth.*

Pilot Knob was the site of the Ely Mining District's first copper properties. Discoveries were made in 1877 and ore shipped to Ward, but no major development took place until after the turn of the century. Joseph Giroux and W. A. Clark bought the Pilot Knob claims in 1900 and employed 50 men. In 1902 the Giroux Mill and Reduction Plant was completed, and serious production began. By 1903, Giroux and Clark had organized the Giroux Consolidated Copper Mining Company, which controlled the Alpha, Taylor, Old Glory, and Pilot Knob claims. These were the best prospects in the district. By 1905 the company town of Kimberly, named for millionaire mine owner Peter C. Kimberly, had a population of close to 100, a post office (Erle Morton, postmaster), saloons, boardinghouses, and numerous residences. A big boost to Kimberly came when the Nevada Northern Railway was completed in September 1906, which not only allowed easier access to the camp but also provided a much cheaper mode of ore transportation.

By 1907 Kimberly had grown enough to warrant the formation of the Kimberly School District. In December disaster struck when the Alpha shaft partially collapsed at the 1,050-foot level, and two miners were killed. Three others escaped, but many days passed before they were finally freed. Kimberly continued to grow, as production in the district increased drastically. By 1910 the camp had grown to 200 and had its own newspaper, the *Kimberly News*. On August 27, 1911, the problem-plagued Alpha Mine was once again

Headquarters and miners' housing for the Giroux Consolidated Mining Company at Kimberly. (Special Collections, Library, University of Nevada, Reno)

the scene of tragedy. A major fire ruined the headframe and most of the shaft workings. Before the fire was brought under control, several Kimberly miners had perished.

In 1916, the Consolidated Copper Mines Company (Charles Boynton, president) gained control of the Kimberly District. The company bought the Giroux Consolidated Mine Company, Copper Mines Company, Butte and Ely Copper Company, Chainman Consolidated Copper Company, and New Ely Central Copper Company. A 500-ton flotation mill, named the Giroux, was built in 1916. By the 1920s, Kimberly was an impressive company town, with a population of 500, a new school, a hospital, a Nevada Northern depot, and many other businesses. The Depression slowed production, and the decline continued until 1958 when the Kennecott Corporation bought the Kimberly townsite. The following year, mill and town were dismantled to make room for the Trippi-Veteran open-pit operation. Today only a large hole and mine dumps show that Kimberly ever existed.

Leadville (Irene) (Saw Mill Canyon)

DIRECTIONS: Located 1 mile north of Monte Cristo.

Leadville lived its entire existence during 1887. Initial discoveries were made in the spring. The Winecup Mine was the only producer. A small camp of 25 formed, and in June Sam Liddle laid out a townsite and asked exorbitant sums for building sites. The people of Leadville were determined to make the town the new boomtown of the White Pine District. By July the town had a saloon (run by D. J. Mahoney), a boardinghouse (George Halstead), and a three-room assay office. For Leadville's Fourth of July celebration, the townspeople composed a song to honor the "greatness" of the town:

> Oh, Leadville, the gem of the Mountain
> the home of the brave and free
> The shrine of each miner's devotion
> White Pine offers homage to thee
> Thy treasure makes capital assemble
> when thy ore veins, bright gems, stand in view
> May thy spurs and summits oft tremble
> by the blast of thy miners so true
> The Wine-cup, the Wine-cup, bring hither
> and fill you it up to the brim
> May the laurels we've won never wither
> nor the star of our prosperity grow dim

May the miners united ne'er sever
but each to his station prove true
Here's to Leadville, our home forever
and the wealth it will bring to us, too.

All the publicity in the world couldn't help Leadville. The townspeople's great pride lasted for only a couple of months—until the mines began to close. By August it became clear that the ore deposits occurred only in very small pockets. The Winecup closed for good in October, and by the first snow Leadville was a ghost town. Because of the poor ore quality and the limited deposits no efforts to revive the district were ever made. Only two mine shafts and some scattered lumber mark the site today.

Lexington Canyon

DIRECTIONS: From Baker, take Nevada 73 south for 4 miles. Exit right and follow this road for 7.5 miles. Exit right and follow for 9 miles to Lexington Canyon.

Silver deposits in Lexington Canyon were discovered in April 1870. Within weeks, several mines were put into production, including the American Eagle, Sunset, Pine Nut, Bald Hornet, Mountain Chief, Blue Cloud,

The Lexington Canyon cabin of three leaseholders—Matson, Harrigan, and Dickson—1917. (USGS)

White Man, and Bob Steel. The assays ranged from $50 to an amazing $16,000 per ton. However, the mines did not have extensive deposits, and within two years most of them had been abandoned. Leaseholders occasionally would work the district, but it wasn't until 1917 that serious mining returned to Lexington Canyon. Tungsten was discovered, and a small mill was built to process the ore. The big producer was the Bonanza Mine, which yielded $20,000 in scheelite before closing in 1918. The district was then idle until 1941, when the Bonanza Mine reopened. A new 50-ton concentrating plant was constructed. The site produced $80,000 before the mine and mill closed for good in 1942. No camp ever formed here, and the only markers are mill ruins and small mine dumps.

Lund

DIRECTIONS: From Ely, take U.S. 6 south and follow for 23.1 miles. Exit left onto Nevada 38 and follow for 11.9 miles to Lund.

Lund grew out of the Tom Plane Ranch, established in 1873. Owned by a mine boss at Ward, the ranch was located on the Hamilton-Pioche stage line. The first flour mill in White Pine County was built in Lund in 1881 and run by a man named Withington. The group of first settlers included John L. Whipple, who later established Sunnyside (Nye County). A townsite was platted in 1898 and named after Anthony H. Lund, an early Mormon settler. The Mormon Church would become the driving force behind Lund and nearby Preston. A post office, with Bryant Ashby as postmaster, opened at Lund on August 27, located in the town office building. The following year, an old bunkhouse was converted into Lund's first schoolhouse. Hattie Rapheal served as the schoolteacher and continued to teach here for many years. Stores catering to local ranches opened, the most successful among them the Lund Mercantile Company, run by Dave Gardner, and a country store, run by Whipple and a man named Judd. Surprisingly, it wasn't until 1903 that the first church, a two-story log structure, was built. In 1908, a more substantial 30-foot-by-50-foot cement-block church was built. Henry Mathis and Merlin Peacock built the Pug Mill in 1907, which produced bricks not only for the Lund area but also for some of the nearby mining towns. In 1915 the Lund District was bonded and a new concrete school constructed, which was in turn replaced in 1931 by a new high school. Lund has continued to be a ranching town, never growing very large but maintaining a steady population of about 100. Buildings from Lund's early years remain.

McGill (Smelter) (Axhandle Springs)

DIRECTIONS: From Ely, head north on U.S. 50 for 13.1 miles to McGill.

Long before McGill became the smelting center of White Pine County, the site was a productive ranch. John Cowger established his ranch in 1872 and soon had extensive grain fields. By 1880 he had become the sole owner of the area's water rights. An unsubstantiated rumor has it that Jesse James and his gang ate here while escaping from a sheriff's posse. William Neil McGill and his partner, William Lyons, bought Cowger's ranch in 1886, and soon the ranch was one of the most prosperous in the county. Lyons had been the co-discoverer of Taylor in 1885, and because of his interests there, he sold out to McGill in late 1886.

A post office opened at McGill Ranch on April 28, 1891, with Kate McGill as postmaster. Five years later, William Neil McGill formed a partnership with Jewett Adams, a former Nevada governor. This powerful union created one of the largest sheep and cattle empires in Nevada. McGill was born in Mount Healthy, Ohio, on January 7, 1853. He worked on the Sutro Tunnel and had surveyed the Ely and Copper Flat townsites. During 1873 he worked for the Martin White Company as a civil engineer. McGill was a man of amazing character. He refused to sell his cattle at a high price to outside interests until the people in Ely, on credit, were supplied. Jewett Adams came to Nevada in 1864 from Vermont. Adams was elected lieutenant governor in 1874 and reelected in 1878. He successfully ran for governor in 1882, although he was defeated in his reelection bid by C. C. Stevenson in 1886. The Adams-McGill Corporation continued for many years, but tragedy struck on June 18, 1920, when Adams died in San Francisco. McGill continued to run the corporation until his death in April 1923. The empire deteriorated rapidly without McGill's guidance, and in 1930 the corporation was liquidated.

However, while the McGill-Adams partnership was important to the town's development, the real force was the McGill Smelter. Construction on the $10 million smelter began in 1906. McGill was selected as the site because the Cumberland and Ely Mining Company owned eight square miles of land, later used for the extensive tailing ponds. The smelter, completed and started on May 15, 1908, was a joint venture of Cumberland and Ely and the Nevada Consolidated Copper Company. A water line was built from Duck Creek to the McGill Smelter, and the McGill depot on the Nevada Northern was completed in 1909.

By 1909, McGill employed 2,200 men. The Steptoe Valley Smelting and Mining Company, operator of the smelter, built large and modern facilities for its employees and to serve as the official company headquarters. The company also opened several businesses, and McGill became a company town.

A view of the McGill Smelter, showing the ore-loading tramway. (Special Collections, Library, University of Nevada, Reno)

The company organized a local newspaper, the *Copper Ore,* as well. The 1920 census showed that McGill had 2,850 people, making it the largest town in White Pine County. The smelter was treating as much as 15,000 tons of ore daily. On July 9, 1922, the nine-acre concentrator burned, and damage to the facility was over $2 million. The mill was quickly rebuilt, and McGill continued to grow.

McGill's population peaked at around 3,000 during the late 1920s and early 1930s. The population declined after the Depression, and by the time the smelter closed in the late 1970s, only about 1,000 residents were left. The smelter was shut down not because of an ore shortage but because it could not comply with strict new EPA air standards without installing extremely expensive air scrubbers. Today the plant has been dismantled. The tall smokestack was taken down in September, 1993. The loss of the smelter had a very detrimental effect, and the town now retains only about 250 residents. Interesting buildings exist there, and a visit is well worth the traveler's time.

Melvin

DIRECTIONS: *From Blaine, continue north for 10 miles. Then make a very sharp right (heading south) and follow for 0.5 mile to Monte Neva hot springs, site of Melvin.*

Melvin was a small camp that sprang up with the advent of the Nevada Northern Railroad. John Melvin built a small ranch at Monte Neva Hot Springs in early 1907. A post office, with Melvin as postmaster, opened on

March 20, and a townsite was platted. The post office served nearby ranches and also the mining camp of Ruby Hill, across Steptoe Valley. Most of the freight destined for that camp came through the Melvin railroad siding. The post office closed on January 31, 1913, but Melvin continued as a successful ranch. The townsite was never needed, since Melvin didn't expand beyond the small-ranch stage. Today Melvin is one of many active ranches in Steptoe Valley.

Menken (Sunnyside)

DIRECTIONS: *Inaccessible by vehicle and almost so by foot. Located on the side of Treasure Hill, 1 mile above Eberhardt.*

Menken, or Sunnyside, was one of many small satellite camps that sprang up during the Hamilton boom. Menken grew up around the Mazeppa Mine, one of the earliest discoveries in the district. It was January 1869 when the Mazeppa Mine was discovered and a rush developed, bringing 300 people to the area. A third of them stayed, and a townsite was laid out in early February. The first official Menken town meeting took place on February 16 at the camp's only store, owned by William C. Masten. On July 16, Edward Applegarth, who already controlled other properties in the Hamilton District, purchased the Mazeppa Mine. In August the mine was producing ore valued at $1,000 per ton. However, by fall the strike had played out, and the camp was abandoned by the first snow. A new mine, the Sunnyside, was worked in the 1870s, but most of the miners lived in nearby Eberhardt rather than in the rougher terrain near the mine. Only mine dumps mark the site today.

Mineral City (Lane City) (Robinson Canyon)

DIRECTIONS: *From Ely, take U.S. 50 west for 3 miles to Mineral City.*

Indian John, a local Indian, took a prospecting party led by Thomas Robinson to silver-bearing ore in November 1867. The Robinson Mining District was organized on March 16, 1868. Within a year, more than 1,200 mining claims had been filed. The most promising mines of the district were the Old England (located December 27, 1867, assayed $185 per ton), Elijah (March 6, 1868, $40 per ton), Springfield (June 1, 1869, $81 per ton) and City of London. A small rush to the area developed, and by 1870 more than 250 people were living in the district. The Mineral City post office opened on August 9 (Henry Hilp, postmaster), and the town began to look quite attractive. By the end of 1870, Mineral City had become the milling center for most of the surrounding districts. The 10-stamp Mineral City Mill had begun production in November 1869, and in early 1870 the Cummings Company built a large smelting furnace. Mineral City reached its peak in 1872–73 with close to 600 residents. The town's businesses included six saloons, four boardinghouses, and several mercantile stores.

The Canton Mining Company, with G. M. Odgen as superintendent, moved into the district. The company controlled 41 ledges, ran a large furnace and smelting works, and employed 60. The best of the Canton mines included the Minora, Blackstone, Yellowstone, New York, Springfield, El Dorado, Randolph, Hayes Extension, and Aultman. In August 1873 the old Piermont Mill was moved here by the Watson Mining Company and put into production. Sagging ore values, however, brought Mineral City's rapid growth to a grinding halt, and by 1874, the population had dropped to 200. Nevertheless, many of the businesses were still able to operate. The Hilp brothers, Henry and Fred, opened a mercantile store that year, and other active businesses included a hotel/restaurant (owned by Fanny Yates), three saloons (all run by Sam Jones), a saloon/hotel (Mace Storer), and a men's furnishing store. In late 1874 Tom Harrigan, Abe Travis, Billy Boyce, and B. F. Miller discovered a new ore pocket in the Elijah Mine and brought some short-lived attention back to the Robinson District. The ore showed 80 ounces of silver per ton and was processed at the Canton Company's smelter. The district continued to fade, however, and by 1880 all of the mines were idle.

A few mines (the Ontario, Arthur, Great Western, West Aultman, and Roadside) were reopened in 1886, but it wasn't until 1896 that Mineral City experienced a big revival. Charles D. Lane, a wealthy Eastern businessman, bought the Chainman Mine and many undeveloped claims. He spent $168,000 to reopen the 10-stamp mill at the Chainman Mine and constructed a power plant and water ditch. Much of the ore, however, was smelted by the Canton smelter. It is interesting that future president William McKinley invested $80,000 in the Robinson District. The Canton Company, based in McKinley's home state of Ohio, helped pay for the favorite son's election in 1896.

In 1900 the Chainman Mine was sold to New York and Pennsylvania capitalists for $150,000. Lane's 10-stamp mill closed in 1901, but a new $100,000, 100-ton cyanide mill was put into operation in March 1902. The new mill, however, was worthless because ore from the district had high levels of copper sulfides, which negated the cyanide process. In 1906 the Chainman, Joana, and other local mines were consolidated into the Chainman Consolidated Copper Company, but the revival was already folding, and by 1910 the district's mines and mills were silent. The post office, which had reopened in 1902, closed for good on July 31, 1911. Because the nearby mines at Kimberly and Ruth have been active through the years, a number of people have continued to live here. Wood and rock buildings, including the school, remain, although only a couple of them were built before the turn of the century.

Minerva

DIRECTIONS: Located 1.5 miles below Shoshone.

Initial discoveries at Minerva were made in 1884 by Orsen Hudson, who opened the first tungsten mines in the district. The Minerva Mining District was formed the following year in September. The three main mines were the Mammoth, the Mooney and Hudson, and the Blue Belle. Heavy mining activity, however, did not begin until 1916, when the Minerva Tungsten Corporation became active. The company controlled seventeen claims, covering 340 acres and containing more than 3,000 feet of workings. A 150-ton concentrator was built, and a 3.5-mile water pipeline to the mill was constructed. The company also developed a camp for the mill's 50 employees. Close to 200 men were employed in the mines and the mill combined. The main producer for the Minerva Corporation was the Chief Mine, located just north of the mill. Activity was suspended in 1918 and the concentrating mill dismantled in 1923. Minerva had a revival beginning in 1936 when the Tungsten Metals Corporation reopened the mines and built a 200-ton mill. Soon the camp had a population of 60. The main producers during the revival were the Hilltop, Tungsten Queen, Everit, Oriole, Chief, and Silver Belle mines. The company abandoned the district in 1940 after it produced $1.7 million. Intermittent activity went on until 1962, but nothing substantial came of it. Today a couple of people still live in the few buildings left. Extensive mine dumps and wooden rubble abound at the site.

Monte Cristo

DIRECTIONS: From Illipah, head west on U.S. 50 for 12 miles. Exit left and follow for 3.5 miles. Bear left and continue for 6 miles. Exit left and follow for 3 miles to Monte Cristo.

Monte Cristo was one of the first camps in the White Pine District, having been organized in the fall of 1865. The Monte Cristo Mining Company, with Thomas Murphy as president, controlled major mining claims in the district. The most valuable mine was the Hidden Treasure #1, but it wasn't until spring 1868 that a serious rush to the Monte Cristo camp took place. During the summer, J. W. Crawford built the 5-stamp Monte Cristo Mill. The first order of business was the production of four silver bars from ore extracted from the Hidden Treasure Mine #2, located at Eberhardt. The population increased to 150, and the town was the site of the polls for the 1868 district elections. To meet the demand for building materials, the San Francisco Sawmill Company built a sawmill to provide a cheap supply of lumber. In 1869 5 stamps, brought from La Plata (Churchill County), were added to the Monte

Only the stack of the Monte Cristo Mill remains today.

Cristo Mill. The White Pine rush, however, focused on Hamilton and Treasure Hill during 1869, and many residents left for greener pastures. The mill closed down, and only minor production remained.

In February 1870 the mill was reopened under the management of H. D. Fairfield. Two Stetefeldt furnaces were added to roast ore from the Mount Ophir Mine. The mill was enlarged again in 1871 to 20 stamps. The Monte Cristo Milling and Mining Company, based in Philadelphia, controlled the Trench, Bald Eagle, and Badger State mines. The Trench was the richest, with ore assaying at $325 per ton. The activity was short-lived, however, for the mill and mines all closed in fall 1872, and within a year the district was abandoned. No revivals took place until the early 1980s, when the Phillips Petroleum Company began molybdenum exploration in the district. Actual mining was begun, and if the initial reports hold true, the Monte Cristo District will be producing for quite a while. However, the drop in mineral prices has slowed current development. The most fascinating of the remains at Monte Cristo is the old smelter stack, which dominates the site. Other stone ruins do exist there, but some of them are behind the fences of the Phillips Petroleum Company. Nevertheless, Monte Cristo is well worth the long trip.

Mosiers

DIRECTIONS: From Ely, take U.S. 50 north for 3.5 miles. Exit right and follow for 7 miles to Mosiers, ignoring all turnoffs to the left.

Sam Mosier came to White Pine in the early 1870s, patented a large tract of land to the east of Ely, and Mosiers Ranch was born. Mosier gained renown as a breeder of fine racehorses. He owned two horses, Harry and Sooner, which were big winners at racetracks in Reno and California. Mosier served as a White Pine County commissioner during the 1870s. When Mosier died suddenly, his ranch was purchased by Henry Hilp of Mineral City. The ranch has had many owners over the years and was active until recent times. Some minor mining activity took place in Mosier Canyon around the turn of the century, when 2,000 tons of fire clay were produced from two large pits in the canyon. Interest in the mines soon faded, however, and the Mosier Canyon Mining District ceased to exist. Today a number of crumbling ranch buildings remain at Mosiers.

Mountain Spring

DIRECTIONS: From Ruby Valley Station, follow Pony Express route east for 9 miles to Mountain Spring. Rough road and rugged driving.

Mountain Spring served as a minor Pony Express station during the last few months of the line's existence. A small stone station was built, primarily as a horse-changing stop. Lafayette ("Bolly") Bolwinkle was the station tender until Overland Stage finally abandoned the stop in 1869. Nothing but scattered stones remain of the original station, located two miles below a corral and woodshed.

The only remains at Mountain Spring. (Nevada State Museum)

Muncy

DIRECTIONS: From Cleveland Ranch, continue north for 23.5 miles. Exit left and follow for 2 miles to reach Muncy Creek Mining District.

N. C. Noe and Frank Bassett, who had crossed the United States in 1860 by horseback, settled and planted a large orchard in the early 1860s. Noe discovered some ore deposits and worked nearby claims, including the Grand Deposit (copper), Kansas (silver), and Blue Hen 1, 2, 3 (copper) mines. In 1870 Fred Plageman, a resident of Eberhardt and Ruby Hill (Eureka County), discovered new copper deposits. The Muncy Creek Consolidated Mining Company was formed in 1872 and gained control of most of the district's mines. The most prominent were the Amargosa, Grand Deposit, Defiance (Cameron), and Kansas (Pioche-Kansas). A small camp of 20 people formed, and Muncy Creek showed promise of being a good copper source.

A post office opened on July 24, 1882, with Jacob Cameron, discoverer of the Cameron Mine, as postmaster. In June 1886, after Noe and Bassett had located a new deposit in February, with values as high as 500 ounces of silver to the ton, a city site, named Defiance, was laid out. The town plan included a courthouse, extensive depot grounds (Defiance was supposed to be a stop on the proposed Salt Lake and Los Angeles Railroad), and a plaza (complete with fountains and pavilion) for the Spring Valley Cornet Band. Noe built and ran a brewery during Muncy Creek's June boom. By August, however, the Defiance townsite was defunct. But production continued, and three new mines were opened: the Lone Cedar (owned by J. M. Ogden of Mineral City), Kate Alice (Lamb, Elmore, and Caruthers), and Centennial.

In 1892, the Silver Mountain Mining Company (J. L. Pressnell, superintendent) bought most of the mines in the Muncy Creek District. In addition to the Defiance and Kansas mines, two new mines, the Keystone and the 76, were put into production. Muncy Creek's fortunes fell during the mid-1890s, and the post office closed on March 21, 1898. A small revival took place from 1908 to 1912 when the Cambria Copper Company worked four claims, known as the Texas group. As a result, the post office reopened on February 20, 1907, and served the small camp until April 22, 1911. Cambria Copper owned water rights along ten miles of Muncy Creek and also controlled 580 acres of nearby ranchland. Ore from the mines was sent to a Salt Lake smelter where it showed 25 percent copper. The company went bankrupt in early 1913, however, after the shallow ore deposits gave out. The property was assessed at $6,600 and sold at a sheriff's auction in July to pay back taxes. Muncy Creek's last revival began in 1919 when C. D. Roy formed the Muncy Creek Mining Company. Roy gained control of 23 claims and the Muncy Creek Ranch, and he also laid out a millsite. The Grand Deposit Mine was the focus of mining operations, and by 1920 a new 25-ton mill and calcinating plant had been put into operation. Ten men were employed, and although the ore was not as rich as earlier deposits, the company enjoyed modest profits for a number of years. In late 1920 the company ran into problems with claim jumpers and finally brought a successful suit against them to get the title to the land. The mining revival died in 1923 and the district returned to ranching. In October 1926 James Miley shot William Cooper, the Piermont Mine boss, who was staying at the Clark residence, apparently because of a claim dispute in the Piermont District. The last mining activity in the Muncy Creek District took place in 1937 when leaseholders reopened the Grand Deposit Mine. A little over $12,000 was removed before the district was abandoned for good. The small ranches along Muncy Creek have continued to operate throughout the years. Only the mill foundation and some scattered rubble now mark the Muncy townsite.

Newark

DIRECTIONS: *From Cold Creek, head south for 21 miles. Exit right and follow for 1 mile to Newark.*

Two prospectors from Austin, Stephen and John Beard, ventured into unexplored Newark Valley in October 1866. They located silver claims later bought by the Centenary Silver Company in 1867. The company spent $50,000 to move a 20-stamp amalgamation mill from Kingston (Lander County) and rebuild it at the newly formed camp of Newark. Another $130,000 was spent adding eight reverberatory roasting furnaces, ten amalgamating pans, five settlers, and a 140-horsepower engine. These additions, however, were a waste of money because the ores of the district were free milling and didn't need to go through the amalgamation process. Money for mill expansion was given by the Methodist Church, with the understanding that the company, when profits rose, would build a new Methodist church in Austin. The main mines of the district (Baystate, Nevada, Battery, Lincoln and Buckeye State) were located in nearby Chihuahua Canyon.

The Centenary Mill handled large amounts of ore from Treasure Hill during the late 1860s when ore production was too much for White Pine mills to handle. Centenary treated ore that was primarily from the Aurora South and Eberhardt mines. A fire struck the mill on February 5, 1868, and while the machinery was barely damaged, more than $20,000 in structural damage did occur. The mill was repaired and back in operation by May 1. The Newark Mining Company bought out the Centenary Company in 1872, and the town of Newark peaked soon after. Newark Mining spent $400,000 on mine development and opened new mines, including the North Chihuahua, South Chihuahua, Washington, and Indian Jim. The summer of 1874 saw close to 200 miners working the district, but the shallow deposits soon ran out and the Newark District began a quick slide to oblivion. The mill closed down in 1874 and was idle through the rest of the decade. Caretakers kept the mill in running condition, and only a few residents stayed in Newark. It wasn't until 1902 that a revival began, with the Newark Mining and Milling Company working several old mines. The company was active until 1908, when ore values dropped enough to force the company to fold. Intermittent activity took place until 1957. Since then the district has been abandoned. Total production stands at $518,000, with the Baystate Mine, at $134,000, being the biggest producer. Considering the money put into developing the district, the returns were not that significant. The mill foundations dominate the site, with mine dumps and collapsed wooden buildings scattered nearby.

Osceola

DIRECTIONS: From Ely, take U.S. 50 south for 34 miles. Bear right off of U.S. 50 and continue for 3 miles to Osceola.

Initial discoveries in the Osceola District were made in August 1872 by Joseph Watson and Frank Hicks, who located the Osceola Ledge. A small camp, named for a Seminole chief, soon formed. The Osceola Mining District was organized in October, and the producing mines included the Exchange (discovered in August by James Matteson), Cumberland (Richardson and Delmater), Crescent and Eagle (Gilmer and Chandler), Verdi (Akey and Felsenthal), Gilded Age (Phillips and Watson), Stemwinder (Charles Bussey

Main Street, Osceola. (Nevada Historical Society)

Osceola at its peak, during the 1880s. (Bancroft Library)

and L. S. Scott), Grandfather Snide (Jack Henderson), Royal Flush (Akey and Delmater), Saturday Night (B. Tilford), and Red Monster (Pat Revey). Osceola grew quickly, and during the camp's first two years of production $300,000 was sent throughout the West via Wells-Fargo.

The placer fields, which put Osceola on the map, were discovered by John Versan in early 1877. Versan sold his holdings to a newly formed company, the Osceola Gravel Mining Company, later renamed the Osceola Placer Mining Company. Soon a tent camp of between 400 and 600 sprang up, and more than 300 miners were employed in the district. The biggest problem was the lack of water to work the placers, which hindered the development of the camp. Osceola nevertheless received national attention when the largest gold nugget ever discovered in Nevada (valued at $6,000) was found here in May 1877 by a man named Darling. In 1878, a small 5-stamp mill was built at the Gilded Age Mine. Osceola continued to grow, and a post office, with Richard Iregaskis as postmaster, opened on March 26. By 1879 the camp had three saloons, a store, a restaurant, a butcher shop, a blacksmith shop, and a justice of the peace.

The lack of water prevented full development of the vast placer fields. Ben Hampton, the head of Osceola Placer, spent $250,000 to construct two ditches to bring water to the district. Hampton set up the first hydraulicing operation in Nevada, and the process found great success at the Osceola placer fields. The company employed 300 men and was active until the turn of the century. During the 1880s, three quartz-crushing mills, of 5, 10, and 20 stamps were active in Old Mill Gulch near Sacramento Summit. The Osceola Mining Company (which owned the Verdi, Mazeppa, Virginia, Durango,

The Nicholson Mill is slowly crumbling.

Sperange, and Ohio mines) started a 20-stamp mill on December 4, 1883, but this mill, like the others, was a dismal failure and operated only on a part-time basis. While mines were active in the district, placer operations were the main producers. Osceola Placer built an electric light plant, office buildings, and homes for personnel on nearby Cemetery Hill. Osceola was one of the few Nevada towns at the time to have electric power, and it also had the honor of having the first telephone in White Pine County.

In 1886 the White Pine Stage Company set up a line from Eureka to Osceola to help keep up with the demand for goods. A fire struck the company's stables on October 31, and ten horses perished, at a loss of $2,000. By 1889, 220 men were still working for Osceola Placer, at wages of $3 per day for powder handlers and $2.50 per day for shovelers. But a fire on April 30, 1890, that destroyed most buildings on the north side of Main Street was the start of Osceola's long, gradual decline. Although activity continued, placer deposits slowly began to run out. This, combined with the deterioration of the water ditches, forced the Osceola Placer Mining Company to fold in 1900. Osceola was not dead though: Many leaseholders continued to work the district, and close to 100 people still lived in Osceola at the turn of the century.

The post office closed on December 15, 1920, but enough people still lived here to support a store and two saloons. The largest revival began in 1925 when the Nicholson Mining and Milling Company began work on seventeen claims, including the Crescent Mine. The 1,500 feet of mine workings yielded ore that assayed $40 to $60 per ton. A 3,500-foot water pipeline was constructed to bring water to an 80-ton mill that the company built in 1927. The company was active until 1932, and leasing activity in the district continued until the late 1950s. Total production for the district stands at $3.3 million, $1.9 from placer mining and $1.4 from lode mining. Until a fire in the late 1950s destroyed all remaining buildings, Osceola always had a few residents. The fire brought an end to the longest-lived placer town in Nevada. Today stone ruins and the Nicholson Mill mark the site. The Osceola Cemetery, on the hill overlooking the townsite, is quite extensive, with the graves of people from all over Nevada. The cemetery is certainly the highlight of a visit to the site. Though it is sad to read the stories on the gravestones, they do reveal the interesting history of Osceola and the surrounding area. Osceola is a fascinating place to explore, well worth an entire day.

Parker Station

DIRECTIONS: *From Ward, backtrack 3 miles to main road. Exit right and follow for 26 miles to Parker Station.*

Parker Station was established in the early 1870s by George Parker, a settler from Philadelphia. He built a two-story log station near Bullwhack Summit that was used as an exchange station for stage, mail, and freight teams. The station was an important stop on the Toano-Pioche stage line. Nevada governor Reinhold Sadler purchased the station from Parker near the turn of the century and lived here on and off until the Riordan family acquired the property in 1917. Today the station is still partially standing, but it does not appear that the structure will survive too many more winters.

Piermont

DIRECTIONS: *From Yelland, continue north for 10 miles. Exit left and follow for 2 miles to Piermont.*

Initial discoveries in the Piermont District were made on July 5, 1869. While ore was of fair value (five tons of Piermont Mine ore returned $337), it wasn't until 1870 that actual production began. In July a new silver pocket was discovered in the Piermont Mine, and ore treated at the Big Smoky Mine returned $180 per ton. The mine, owned by Laughlin, Godfrey, and Wilson, was offered for sale at a price of $60,000, and the newly formed Piermont Mining and Milling Company purchased the property. The company, with J. E. Perkins as president, employed 30 in the Piermont Mine, and in May 1871, put a 10-stamp mill into production. A rush to the district

Only the solid cement foundations of the Piermont Mill remain.

began in earnest during the summer and by the next year close to 400 people were crowded into the canyon. During the first four months of 1872, close to $20,000 was produced. Then by 1873 the boom went bust. Both mill and mine closed during the spring, and by the summer the town was virtually abandoned. The mill was sold in June 1873 to the Hayes Mining Company and was moved to Mineral City.

Except for occasional leasing activity, the Piermont District was abandoned until the early teens, when the Glendale Mining Company was organized. It wasn't until 1920, when Glendale was renamed the Piermont Mines Company, that serious mining activity began. The Cocomongo Mill of Egan Canyon was moved here in 1922. In Piermont's early days, ore from the mines had to be hauled 160 miles to Austin for treatment. Now the treated ore had to be shipped only 27 miles to the Ray siding on the Nevada Northern Railway. In 1924, the Ely-Calumet Leasing Company leased the Piermont property and immediately built a diesel electric plant and a mill, plus a 10,800-foot water pipeline. Two bunkhouses, a boardinghouse, and a log cabin left from the 1870s were renovated, and eight new homes were built. The "floating" Aurum post office was at Piermont for a while, with Addie Dolan serving as postmistress.

More than 100 men were employed, and electric locomotives hauled the ore from the mine to the mill. Piermont was not without some local controversy, however. The old Piermont graveyard, with about a dozen graves, was located just below the mill. The company wanted to dump ore on the site but Piermont old-timers protested. Progress won out, however. All the grave markers were removed, and the ore was dumped. The markers, minus their namesakes, were placed in a new cemetery. In 1928 Piermont Mines, Inc., was organized by E. W. and W. H. Venable and gained control of four other companies in the district—the Ely-Calumet Mining Corporation, the Ely-Calumet Copper Company, the Piermont Mines Company, and the Ruby Hill Apex Mining Company. The company was active until 1936, although the mines and mill were operated only intermittently after 1930. Total production for the district was more than $2.5 million. Piermont Canyon is rich with ruins of all sorts. Concrete foundations from the 1920s abound at the site, and wooden buildings, survivors of activity before the turn of the century, also remain. The "new" graveyard is marked by wooden headboards located at the base of an ore dump. Piermont is a delightful place to visit, and the traveler can happily spend a full day here exploring the site. A beautiful stream flows through the site and makes the area an excellent camping spot.

Pinto

DIRECTIONS: *Located 5 miles southeast of Eureka, on U.S. 50.*

Discoveries were made at Pinto as early as 1865 by Moses Wilson, but it was not until March 20, 1869, that the Silverado Mining District was organized by a person named Duquette. Pinto was the scene of frenzied smelting activity as demand for smelting facilities for the Eureka Mines rose. Despite the emphasis on smelting, Pinto also had several prominent mines, including the Champion (December 1869, purchased in July 1870 by the Nevada Land and Mining Company), Maryland (July 1869, sold for $300,000 in March 1871), Mountain Chief (March 1869), and Wisconsin (July 1869). Other producers included the Compromise, Robert E. Lee, Cole and Johnson, American Eagle, Rescue, and Queen.

The town of Pinto grew as fast as the mining and smelting did. During 1869 the town had a population of about 100, four stores, three saloons, a couple of boardinghouses, and a few other businesses. A school was also built, as was a three-story Mason and Odd Fellows hall. Lots on Main Street sold for as much as $2,000. Two sawmills were built nearby to supply the town with building supplies. A post office, with James Frank as postmaster, opened on September 9, 1870. But by this time Pinto was past its prime. The ore deposits proved shallow, and that, combined with the construction of huge smelters in Eureka, spelled doom for the promising town of Pinto.

Virtually all the mills and smelters were idle by April 1870. A small amount of mining activity continued, keeping about 50 residents in the town. The post office closed on February 15, 1871, but the Pinto Silver Mining Company continued to produce ore and run its smelters. The company smelted $22,000 worth of silver in 1872 and spent $100,000 on a 540-foot connecting tunnel to the Maryland Mine. However, the company folded the next year when no new ore was found in the Maryland Mine. The town struggled to survive, and even though the mines and mills were closed, about 25 people still resided here in 1880. By 1884, however, Pinto had joined the ghosts. The district experienced only a minor revival when some intermittent activity took place from 1908 to 1922. Total production for the district was $226,500. A few stone walls and mill foundations are what is left to mark the Pinto townsite.

Pinto Creek Station (Sharmen's Station)

DIRECTIONS: From eighteen-mile house on U.S. 50, head north for 6 miles. Exit left and follow for 2 miles to Pinto Creek Station.

Pinto Creek Station was built in the late 1860s and served as a stop on the Austin-Hamilton stage line. John McQuig visited the station on July 10, 1869, and reported that there were no accommodations and only a few prospectors at the station. Pinto Creek Station was the scene of a violent murder in 1873, when a man named Green was killed by multiple blows from an ax. The two killers, Dan Matheny and a horse thief named Finney, stashed Green's body under a nearby bridge. Finney left seventeen-year-old Matheny to take the rap for murder, and he was sentenced to hang at Hamilton. A reprieve was granted, and Matheny was sent to prison in Carson City. Finney was never caught. A small ranching settlement developed around Pinto Creek Station during the 1870s. Despite the closing of the stage line, Pinto Creek had a population of 29 in 1881. Today the area is an active ranching district, and period buildings remain.

Pogue's Station (Pancake)

DIRECTIONS: From Eureka, head east on U.S. 50 for 10 miles. Exit right and follow good road for 8.5 miles. Exit left and follow for 7.5 miles to Pogue's Station.

Pogue's Station was located on Pritchard's fast freight route, which ran from Palisade to Pioche. Nearby springs were the only source of water for miles. Stations were set up on the stage line every 25 miles. Pogue's Station was named for and run by Old Jim Pogue, a Kentuckian. Pogue fit the image of an old-time mountain man and enjoyed portraying that image. The station itself was not fancy, but it was a substantial adobe structure. Within a year after the station was built in 1871, Pogue had built a barn and a complete complement of corrals. After Pogue died in 1915, rumors ran wild that he had buried a fortune in gold coins nearby. Over the years, treasure hunters have dug holes all over the site and destroyed the old station, all to no avail. Maybe the gold never existed—maybe it is still there. Only faint adobe and stone walls now mark the remote site.

Preston

DIRECTIONS: *Located 5 miles north of Lund.*

The town of Preston grew out of a Mormon settlement founded at the Maddox Ranch in 1876. Preston was named after William B. Preston, the fourth presiding bishop of the Mormons. The Mormon Church would become the moving force behind the development of Preston and nearby Lund. The settlement began to grow rapidly after 1898, and the first frame house built belonged to T. D. Bradley. Soon after, the town's first store, run by Mort Peterson and H. A. Comins was opened. A post office, with Oliver Cloward as postmaster, opened on September 7, 1899. After the turn of the century, a sawmill (1900), a combination social hall–school–church (1903), and a four-room cement school (1915) were built. Preston has maintained a serene existence as a ranching community through the decades. The post office closed on August 31, 1952, but Preston still has a stable population of about 50. Several buildings from Preston's beginnings remain.

Regan

DIRECTIONS: *From Claytons, continue east for 4 miles. Exit left and follow for 0.75 mile. Exit left (south) and follow for 3.75 miles to Regan.*

Regan was a short-lived mining camp that came into existence during the summer of 1906. Tungsten ore was discovered by David Regan, and a small camp of 20 people formed. A post office opened at the camp on August 20, with William Henriod as postmaster, but after the mines closed, the post office followed suit on November 30, 1907. The office was moved to Trout Creek, Utah. A small revival took place in 1910 when C. G. Simms and Casten Olsen dug a 200-foot tunnel and several smaller shafts. After limited production, this activity ceased. Today nothing remains at the site except some small scattered mine dumps.

Reipetown (Reifetown)

DIRECTIONS: Located 2 miles southeast of Ruth.

The Reipetown townsite was platted by Richard Reipe, who had located the nearby Garnet Mine, on December 6, 1907. Reipetown was designed as an alternative to Kimberly and Ruth for the district's miners. By spring 1908 Reipetown already had a dozen saloons and was well on its way to becoming the "wettest" town in White Pine County. Known as one of the roughest towns in the state, Reipetown became a haven for liquor, gambling, and prostitution. Knifings, fights, and robberies became commonplace there. Reipetown has the distinction of being one of few Nevada towns never to have had a church! A fire struck during the summer of 1908, destroying the Mint Saloon and four other buildings. This potential disaster did not seem to slow the town's progress: By spring 1909 close to twenty saloons were in operation. A post office was organized on May 1 and served the town until April 30, 1912. Reipetown achieved its peak population of 200 during 1909 and maintained that population until the fire of 1917, which not only wiped out the saloon and red-light district but left only two houses unscathed. The town rebuilt quickly but never recaptured its past glory. White Pine County passed strict controls on liquor licenses in late 1917, and so Reipetown was forced to incorporate on January 27, 1918, which allowed the town to issue its own liquor licenses. Needless to say, no requests were denied! The state was not appreciative and forced the town to disincorporate in 1919. The passage of the Eighteenth Amendment dried out the town to some extent, but Reipetown had more than its share of bootleg bars. Fire struck again in 1924 and wiped out all the saloons on the north side of Main Street. The town has slowly faded through the years and is no longer the boisterous place it once was. Today only a handful of people live there. Buildings, including many old shacks, still remain at Reipetown, and the residents hope that if the nearby Ruth copper mine is reopened, the town will revive once again.

Round Springs

DIRECTIONS: From Illipah, take U.S. 50 east for 4 miles. Exit left and follow for 2 miles to Round Springs.

Round Springs was settled in the 1870s by George Halstead from Utah. He built an inn here to serve travelers on the Ely-Hamilton stage. Halstead also made Round Springs into a toll station, setting up a gate near his inn and charging a small fee for passage. In the early 1880s he was bought out by Amasa Lyman Parker, a sheepman from Utah who had several thousand sheep. Despite the harsh feelings that cattlemen had toward sheepmen, Parker became one of the few respected sheepherders in the White Pine area. He was very careful about trespass and was held in high regard. Parker formed a fast friendship with A. C. Cleveland, a powerful local cattleman, was appointed to a county commissioner post, and was considered the safest and most trustworthy county official. In 1896 Jewett Adams and William Neil McGill, also prominent White Pine cattlemen, came to Parker to buy the first sheep for their ranching empires. The two men bought two thirds of Parker's herd, and the following year Parker sold his ranch, land, and remaining sheep to McGill and Adams. The Round Springs Ranch was used for years but was eventually abandoned, although the range is still used by local cattlemen. Today little is left of Round Springs. Only foundations, a small shed, and an old corral mark the site.

Ruby Hill (Medina) (Rubyville)

DIRECTIONS: *From Kent, head north for 3.5 miles to Ruby Hill.*

Little has been written about Ruby Hill, but this small mining camp was active for many years and produced close to $200,000. The initial location, the Cow and Calf Mine, was made by Tom Andrews on July 16, 1871. Within weeks a small camp formed, and prospectors soon had the area claimed. The Ruby Hill Mining District was officially organized in January 1872 by William Adams and Richard Whitworth. Active mines in the district included the Cow and Calf, Silver Wreath, Lookout, Northern Light, Wide West, Gallagher, Monitor, Rattler's Joy, Grizzly, and Iowa Chief. A 5-stamp mill, the Henderson, was started in January on Cow and Calf ore. The Ruby Mining and Milling Company owned the mill, as well as other promising mines in the district. Unfortunately, the mill ran for only a short while because the ore had too much antimony for successful amalgamation.

In May a new mine, the Birch, was put into production. Its ore was rich enough that the Birch Silver Mining Company, run by Bird, Fitzhugh, and Taylor, completed a new 5-stamp mill in August. Also by August Ruby Hill reached its peak, with a population of 150. Two restaurants, two stores, a bar, and a boardinghouse made up the camp's business district. Freight was brought in from Toano (Elko County) at a cost of $30 per ton. By the fall, all the smaller mines had played out, and only four large mines were still producing: the Northern Light (Taylor, Leach, and Company), Bay State (Bird and Company), Cow and Calf (Whitworth and Adams), and Birch (Bird, Fitzhugh, and Taylor).

In 1873 all the mines in the Ruby Hill District were bonded to Colonel O'Connor Sidney. His involvement with English capitalists resulted in extensive litigation, which closed the district's mines and emptied the once-bustling camp. It wasn't until the early 1880s that some mines were reopened.

W. B. Lawlor, head of the Oregon Mining Company, owned the richest and best claims in the district. Ore was shipped to the Aurum Mill. By 1885, however, Lawlor and his wife were the only residents left in Ruby Hill. In 1886 B. F. Brooks of Boston gained control of a couple of mines, and Ruby Hill soon had a population of 15. In January 1887 Brooks purchased the Silver Wreath, Cow and Calf, and Lookout mines for $14,500, from the three men who owned all three mines, B. B. Bird, J. B. Williamson, and F. A. McDennis. The mines had been idle for ten years. Brooks, however, died soon after the deal was completed, and battles between the heirs closed the mines again. It wasn't until November 1892, when Lawlor sold the Grizzly Mine to the newly formed Grizzly Mining Company for $40,000, that Ruby Hill became active once more. The company consisted of a group of prospectors from Ely that included Isaac Jennings, J. R. Sharp, L. G. Hardy, A. E. Hyde, John Beck, A. M. Cannon, J. M. Fox, D. W. James, and J. E. Jennings. The company's persistence paid off when a big strike was hit in July 1893, with ore assayed as high as 20,000 ounces of silver per ton. The mine was worked until summer 1895, when it was shut down after producing about $50,000.

Ruby Hill was abandoned until 1923, when Oscar Siegel, the founder of nearby Siegel, reopened a couple of the mines. In 1924 Siegel sold a new mine, the Ruby Hill Annex, to the Ely-Calumet Mining Company. The mine was located high on Ruby Hill and horses were needed to haul ore. All the holdings in the Ruby Hill District were purchased by Piermont Mines, Inc., in October 1926. All mines were located between 8,000 feet and 8,500 feet altitude. The ore was of sufficient value, but the difficulty of removing it and the cost of shipping it to mills forced Piermont to shut down in 1928. This brought to an end all mining activity in the Ruby Hill District. Because the camp was active for only a few years at a time, few permanent buildings were ever constructed. Wooden remains and mill foundations mark the site. Most of the mines are high in the mountains, and access is extremely difficult. Though the ruins are not extensive, Ruby Hill is worth the trip, primarily because of its beautiful setting, nestled high in the spectacular Schellbourne Mountains.

Ruby Valley Station (Ruby Station) (Overland Ranch)

DIRECTIONS: From Fort Ruby, head south for 4 miles. Exit left and follow for 1.75 miles to Ruby Valley Station.

Ruby Valley is among the most beautiful areas of Nevada. Early emigrants traveling through the area named the valley. It wasn't until 1859 that Colonel "Uncle Billy" Rogers set up a trading post and the area was settled. When the Pony Express was organized, Ruby Valley Station was built.

Not only was Ruby Valley an important stop for the riders but a nearby ranch provided the tired horses with some of the best feed along the Nevada portion of the route.

Once the Pony Express was in operation, Rogers built a sturdy stone station house. Rogers was a former El Dorado County, California, sheriff and in 1860 was named an assistant Indian agent for the area. Several impressive Pony Express rides originated from Ruby Valley. Among the most important began on July 3, 1860, when twenty-year-old Frederick Fisher, whose normal run was Ruby Valley to Egan Canyon, rode 300 miles to Salt Lake City in 34 hours to alert troops at Camp Floyd of an Indian uprising. The troops arrived in time to stop the unrest.

A post office opened at Ruby Valley on April 30, 1862, with C. R. Stebbins as postmaster. Flora Bender wrote, "There is a post office, telegraph office and stone station here, besides a number of log cabins." On September 1, 1862, Colonel Patrick Edward Connor arrived with troops to establish an outpost for Indian protection. Fort Ruby was built two miles away, and emigrants traveling through Ruby Valley had to take an oath of allegiance from the soldiers.

Although the Pony Express folded, the arrival of the Overland Stage the following year and the proximity of Fort Ruby kept the station's importance alive. Ira D. Wines, a prominent Elko County citizen, soon became a major part of Ruby Valley as the postmaster, operating out of a three-room log general merchandise store that he owned. Supplies came via Wines's Halleck–Ruby Valley stage line. Wines also acquired the nearby Overland Ranch and constructed a flour mill on Overland Creek. The mill was active until 1907. Wines died in Salt Lake City in April 1923. Once the Overland Stage ceased operations in 1869, Ruby Valley Station ceased to exist, and the area became the widely spread-out ranching community it is today. Now only scattered stones mark the site of the Overland Stage station and a historical marker shows where the Pony Express station once stood. The station itself was removed by the Northeastern Nevada Historical Society in 1960 and now stands in front of the Northeastern Nevada Museum in Elko.

Ruth

DIRECTIONS: From Ely, head west on U.S. 50 for 6.1 miles. Exit left and follow for 2 miles to Ruth.

The town of Ruth formed in 1903, one of several small towns that sprang up around the huge open-pit copper mines. The town was named for the daughter of D. C. McDonald, a local mine owner. The Nevada Consolidated Copper Company built an experimental mill in 1904, and soon after

The Giroux Consolidated Mill at Ruth. (U.S. National Archives)

that the company town of Ruth was built. Buildings constructed included boardinghouses, bunkhouses, and a two-story hospital. A post office opened in February 1904. Ruth actually was a boomtown during 1909 and 1910, with a population of more than 500. Ruth was a quiet town, however, since bars were banned. There was little trouble in the town.

As long as mining was good, Ruth continued to prosper. By 1928 the town had a population of 2,200 and an impressive business district. The two major mining companies in the district were the Consolidated Coppermines and the Kennecott Copper Corporation (formerly Nevada Consolidated). The two were in partnership on most of the Ruth property. In 1958, Kennecott gained complete control of the mines, including the Liberty Pit. The old town of Ruth was moved to a new site to allow for expansion of the open-pit operation. The Liberty Pit was abandoned in 1967, and a new pit began the following year. The mines have been operating only intermittently during the past few years, and as a result Ruth has lost most of its population. Once the mines reopen, Ruth will probably grow again, but the mines probably won't be reopened until prices for copper go high enough to warrant shipping the ore to outside smelters. Nothing remains at old Ruth, now the site of the huge Ruth pit. Most buildings at new Ruth are of recent origin.

The only building remaining at Keystone, located at the junction of U.S. 50 and the road from Ruth.

San Pedro

DIRECTIONS: *From Osceola, continue west for 1.25 miles. Exit left and continue for 1.5 miles. Exit right and follow for 2 miles to San Pedro. Roads are very rough to this site.*

San Pedro was a small, short-lived mining camp that sprang up during the Black Horse boom because of its accessibility to water. Since Black Horse had no running water, San Pedro became the focus of the Black Horse District when a mill was completed in the summer of 1911. A post office opened on November 6, with William Miller as postmaster, to serve the 15 mill workers and their families. However, the Black Horse boom died quickly, and the mill closed down in the summer of 1912. San Pedro soon emptied, and the post office closed on September 30, sealing the camp's fate. Nothing remains at the site today.

Schellbourne (Fort Schellbourne) (Schell Creek Station)

DIRECTIONS: From Cherry Creek, backtrack to U.S. 93, then head south for 6 miles. Exit left and follow for 4 miles to Schellbourne.

Schellbourne has had one of the most intriguing and varied histories of any of White Pine County's ghost towns. The town's illustrious story began in the spring of 1860 when the Pony Express built a station house and corral. The Schell Creek Station was constantly plagued by Indian problems and was struck by a vicious attack on July 16, 1860, by the same Indians that had escaped from Lieutenant Wood at Egan Canyon earlier that day. The attack left the station keeper and two assistants dead and led to the formation of Fort Schellbourne. A small detachment of soldiers was stationed there until summer 1862, when tensions finally subsided. Schell Creek had its share of intriguing Pony Express characters. Most prominent was Elijah Nichols "Uncle Nick" Wilson, who was the first rider assigned to the Schell Creek–Deep Creek run. Wilson had several skirmishes with the Indians, including one in which he was struck in the head by an arrow. He survived, but suffered from headaches for the rest of his life. After the Pony Express folded in October 1861, the Overland Stage took over the station. John Fisher, a former Pony Express rider on the Egan Canyon–Schell Creek run, became stage driver on the Overland for the Schell Creek–Salt Lake run. The Overland stopped running in 1869, but smaller stages continued to run through Schell Creek.

This old stone store, which possibly served as the Wells-Fargo office, still retains its heavy metal doors.

Soon after that, Schellbourne became a mining camp. Initial discoveries were made by James McMahon during spring 1871. McMahon and his two partners, Henry Gilbert and J. D. Holland, formed the McMahon Gold and Silver Mining Company. Twenty-two men were employed at the Woodburn Mine, which had ore that assayed as high as $800 per ton. By fall Schellbourne had a population of 100, and on December 27, a post office opened here with Melchior Rowan as postmaster. In February 1872 a townsite was laid out by R. D. Ferguson. Businesses included five saloons, a Wells-Fargo office, two blacksmith shops, three stores, two restaurants, two lodging houses, and two law firms (Tilford, Hillhouse, and Ferguson; and Thorton, Ashley, and Stonehill). During summer 1872, the Schellbourne boom reached its peak. Close to 400 residents were crowded onto the flatlands around the old Pony Express station. Three small quartz mills were in operation and in July the *Schell Creek Prospect* began publication. Forbes and Pritchard, of the *White Pine News,* set up the paper, which ran only until January 1873. Then all the machinery was transferred to Battle Mountain and used to run *Measure for Measure.* Although the three quartz mills were idle, Schellbourne's future looked rosy in 1873—until rich strikes were made at Cherry Creek, directly across Steptoe Valley. This new production stifled the Schellbourne boom and drained most residents. Many buildings were moved to Cherry Creek, leaving gaping holes in the main street business district. Some mining activity continued during the 1870s, but the boom days were over.

By 1881 Schellbourne still had a population of 85, most of whom were staying in hopes of a revival of the mines. Some Schell Creek mines were reopened. A nearby mill, El Capitan, was refurbished and started in the spring. A school opened on June 6, and five students were taught by Blanche Lewis. Two mines, the Woodburn (formerly the Northern Light) and the Summit, continued to supply the mill with $200-per-ton ore. El Capitan was shut down during summer 1883, but some limited mining activity did continue with very limited production. By 1884 Schellbourne's population had shrunk to 25.

Fire struck the town on July 27, 1884, when the home of William Burke (Schellbourne's most prominent resident) was destroyed. Schellbourne was, however, quite free of disastrous fires. Unfortunately for the town, the early 1880s ended all serious mining activity, forcing virtually all the businesses to leave the town. The post office did remain open until October 15, 1925. Schellbourne has enjoyed a quiet and serene existence for more than a hundred years. While never abandoned, the town hasn't had more than 25 residents since the turn of the century. Today a small ranch operates there, amid some beautiful stone and log buildings. Schellbourne is a definite must-see for the ghost town enthusiast. The photographic possibilities here are endless.

Seligman (Leadville)

DIRECTIONS: From Illipah, head west on U.S. 50 for 12.5 miles. Exit left and follow for 3.5 miles. At fork, bear right and follow for 4 miles. Exit left and follow for 2.5 miles to Seligman.

Seligman was one of the last mining camps to become established in the White Pine Mining District. Discoveries were made in spring 1886. By the summer a Wells-Fargo office, a triweekly post office, a store, a boardinghouse, and a feed stable had opened. The most interesting aspect of Seligman's growth was the absence of gambling houses and saloons. The area's major mines were the Crusader, Dead Broke, and Crusader #2, all owned by the Arm and Hammer Baking Soda Company. The townsite, founded by Eugene Robinson in December 1886, was named Leadville. In July 1887 the town was rechristened Seligman, and the mining district was surveyed by Robinson, an officer in the Sweetwater Mining Company. By August Seligman had grown to 100 residents and had ten houses, an assay office, and 80 men working in the mines. An official post office, with George Halstead as postmaster, opened on September 19, and the camp took on an air of permanence. The mines in the district were rich enough to warrant serious consideration of building a railroad to Eureka. The Seligman, Eureka, and Nevada Southern Railway Company never became a reality because the mine deposits turned out to be quite shallow. During July 1888 close to 200 men were employed in the Seligman mines. The camp enjoyed its peak that summer, but then, one by one, the mines ran out of ore and shut down. By 1890 the population had dwindled to 75, and 1895 saw only 50 residents left. The post office managed to remain open until December 31, 1905. Soon afterwards, Seligman was totally abandoned. Only an occasional prospector has lived in the camp since. Recently some new mining activity has taken place at Seligman. Major plans for the district are in the works, but whether the deposits are large enough for consistent production remains to be seen. Stone ruins and foundations remain at Seligman. The townsite, deep in beautiful Seligman Canyon, is well worth the trip.

Shermantown (Silver Springs)

DIRECTIONS: From Eberhardt, continue south for 1.5 miles. Exit left and follow for 2 miles to Shermantown.

The new camp of Silver Springs was established during the summer of 1868 by Major E. A. Sherman and Joseph Carothers. Because of the camp's favorable location, mills were planned to treat ore from the rich Treasure Hill mines. The *Daily Alta Californian* of November 26 reported that Silver Springs had two or three good brick buildings. It also boasted a sawmill, a quartz mill (10-stamp Oasis), a smelting furnace, and an assay office. By the end of the year, two sawmills (Stanford Hall and A. H. Ruthford) were producing lumber constantly to keep up with the camp's wood demand. The Oasis Mill, originally the Keystone Mill from Austin that had burned in 1867, was the first mill and began operations in October. It was originally called Page's Mill, after its owner. The mill had no roasting or chlorination capabilities, just simple amalgamation. Silver Springs businesses were joined by a two-story, sixteen-room hospital built by Helena Jones of Dayton. Two law firms also opened and the demand for their services was great, for claim jumping was a common occurrence in the White Pine District.

Eighteen hundred sixty-nine was the big boom year for the camp, which was renamed Shermantown in January and became the milling center for the White Pine District. Eight mills and smelters were in operation during the year. The largest was the Metropolitan, which was started in June. The mill, owned by the Metropolitan Silver Mining Company (Alpheus Bull, president), contained 15 stamps, ten amalgamating pans, five settlers and processed 20 tons a day. The Dayton Mill was built next to the Metropolitan. Both mills were heavily damaged by a rare tornado on August 28. The Oasis Mill, owned by the Eberhardt Company, was refurbished early in the year, and four furnaces and five pans were added. The Moore and Barker Mill, formerly the Butte Mill of New York Canyon in Austin, was started in January 1869 and contained eight 650-pound stamps. Other mills included the Kohler, or Staple's (May 12, 1869, processed ore from the Babylon Mines, enlarged to 20 stamps in September), Drake and Applegarth (8 stamps), and Moyle and Sears (5 stamps).

Shermantown's business district also boomed during 1869. Saloons were extremely popular, and among the best were the Headquarters (J. H. Vanderbilt and D. W. Evan), Wilkin's (D. Wilkin), Gem (J. C. Scott), and Oro Fino (Wilson and Thompson). A telegraph line was completed to Shermantown in April, and the Silver Springs Water Company (Luther, Hevy, and Page) channeled water to the camp via a 900-foot tunnel, that provided fourteen inches of water. A post office opened on April 30, and by that summer Shermantown was rumored to have as many as 3,000 residents. Two newspapers also began publication. The *White Pine Evening Telegram,* founded by Edward McElwain and V. E. Allen, started on June 2, with a yearly subscription rate of $16. Allen gave up the business after only a month, and despite McElwain's determination to succeed, the paper folded in November. The *Evening Telegram* just wasn't able to overcome the popularity of the *White Pine News.* The *Shermantown Reporter,* on the other hand, was politically motivated. The paper, organized by Pat Holland, was printed specifically to take subscribers away from the *Inland Empire* and help the *White Pine News.* After the *Inland Empire* folded, there was no need for the *Shermantown Reporter.* Its presses were sent to Eureka for the *Eureka Sentinel* in June 1870.

On June 24, 1869, the cornerstone for the impressive three-story Mason and Odd Fellows hall was laid. Most buildings in Shermantown were built of pink sandstone, quarried locally. Shermantown's glorious bubble burst in 1870. The town's entire existence was inseparably tied to the mines of Treasure Hill, and when these mines faltered in late 1870, Shermantown was doomed. By springtime, only 200 residents remained. The post office closed on June 19, 1871, and Sherman's slide quickened. By 1875, only Dr. E. X. Willard and his family resided in the town. Most buildings were moved to Hamilton after the big fire there in 1873. No revivals have ever taken place here, and after the Willard family left in 1891 the town became a permanent ghost. Today impressive stone mill walls and smelter stacks mark the site. Despite the tricky drive to the site, Shermantown is well worth the trip.

Shoshone

DIRECTIONS: From Ely, take U.S. 50 east for 26.5 miles. Exit right onto U.S. 93 south and follow for 3.9 miles. Exit left and follow for 15.2 miles to Shoshone.

The Shoshone District was formed in 1869 after Benjamin Kimball made initial ore discoveries. The main producer of the district was the Indian Queen Mine. A camp formed in the 1880s, and by 1896, the town was large enough to be honored with a post office, with George Swallow serving

The Shoshone Mill in 1945.

as postmaster. While mining in Swallow Canyon was not very profitable, the huge success of the Minerva mines, located a couple of miles to the south, kept Shoshone alive for many years. Many Minerva miners preferred to live here. The school for the district was built on nearby Shingle Creek and served the populace for many years. The post office closed on August 31, 1959, and today Shoshone has only a handful of people left. Several buildings remain at Shoshone, and additional ruins are located in Swallow Canyon.

Siegel (Centerville) (Queen Springs)

DIRECTIONS: From Stone House, continue south for 3 miles. Exit right and follow for 3.5 miles to Siegel.

Centerville was another of many mining camps that sprang up throughout the Schellbourne Range during the 1870s and 1880s. While these camps never grew to sizeable proportions, their history is an important part of White Pine County's development. Initial discoveries were made by Samuel Jameson and E. G. DeMill during spring 1871, and the Queen Springs Mining District was organized on June 24. A handful of prospectors inhabited the district, but it wasn't until the following summer that the camp of Centerville actually formed. In August 1872 the Tehama Consolidated Silver Mining Company bought Jameson and DeMill's claims for $28,500, and the following month construction was begun on a 20-stamp mill. The company's owners were mining men who had been active in the area: John Biggs, George Treat, J. W. Dickinson, Archibald McDonald, Edgar Mills, W. H. Duren, and James Cameron. The company employed 50 men and started its $75,000 mill in November. The mill, however, failed because there was not enough millable ore. By fall 1872 the camp had a population of about 75, and a 160-acre townsite, owned by R. J. F. McWilliams and Company, was platted.

Production was not spectacular, but many small mines were active in the district, including the Great Western, El Capitan, Snowstorm, Morning Star, Anna, Monitor, Simon, Mohawk, Nantucket, Hidden Treasure, Mahogany, and Argenta. All of the properties were bought by Treat, Biggs, and Richardson during spring 1873, but ore occurred only in surface deposits. By 1874 only a handful of prospectors were left, and after the Tehama Mill was bought by the Chance and Baltic Company and moved to Cherry Creek in April, Centerville was abandoned.

In spring 1881 the camp was revived when Simon Davis made some discoveries. El Capitan Mine was reopened, and a small 5-stamp mill began operations in July, but by the end of the year this revival had faded completely. There was some excitement in 1882, but it had nothing to do with mining: It was in Centerville that Senator Spencer of Alabama hid after revealing the

The last buildings in Siegel, shortly before their removal by the United States Forest Service. (USGS)

Star Route mail fraud. Revenge seekers searched the West for Spencer, but he found safe haven until the perpetrators of the fraud were put in jail.

Centerville's last revival began in 1903, when A. L. Siegel, of Salt Lake City, bought all the claims in the district. The camp was renamed Siegel and became the company camp of the Siegel Consolidated Mines Company. The active mines included the Siegel, the Gold Crown, and the May Queen group. A post office opened at Siegel on January 19, 1907, with Harold Siegel as postmaster. Ore deposits ran out, and by summer 1908 all mining activity had ceased. The post office closed on July 31, and by fall Siegel was abandoned for good. At the time, seven buildings remained, most from the 1870s. All but one of the buildings were removed during the 1930s. Today the ruins of that building, mill ruins, and a couple of stone foundations mark the site. As are all the mining camps located in the Schellbourne Range, the Siegel townsite is in a beautiful setting, and that in itself makes the hard trek worthwhile.

Spring Valley Station (Stone House)

DIRECTIONS: From Cherry Creek, take Nevada 489 to U.S. 93. Exit right and follow for 6 miles. Exit left onto Nevada 2 and follow for 14.8 miles to Spring Valley Station.

Spring Valley Station, or Stone House, was a stop on the Overland Stage route from 1861 until 1869. After the Overland folded, the station was used by the Woodruff and Ennor stage line. A fine stone station house was built, and lodging and fine food were offered to weary travelers. Extensive corrals and a huge horse barn complemented the stately station house. The station tender was Constant Dubril, who delighted in preparing home-cooked meals for her patrons. On April 25, 1870, the station was hit by a fire, which only slightly damaged the station house but destroyed the horse barn, a huge stock of hay and barley, and ten horses that belonged to the Woodruff and Ennor company. The losses were estimated at $4,000. After the stage line stopped running, Stone House was active as a small ranch. Today the beautiful stone house still stands. A unique part of the house is the well, which is located inside the building. Several other period buildings also remain. Spring Valley Station is worth the trip.

Springville (Maquire) (Whitworth's)

DIRECTIONS: *Located 1 mile north of Centerville.*

Springville was a small satellite camp of Siegel. The camp formed during summer 1872 and by August Springville had 25 residents. The Home Ticket Mine was the big producer for the camp and employed 15 men. The Pennsylvania Mining and Milling Company owned the mine, and on August 10 put into production the 5-stamp Livingstone Mill. Production, however, was not sustained, and by the end of 1873 the camp had been abandoned. Only a handful of buildings (one a hotel) were built before the Home Ticket closed. Today only the mill foundations and some scattered rubble mark the site.

Steptoe City

DIRECTIONS: *Located 2 miles northwest of McGill.*

Steptoe City's growth paralleled that of McGill. When McGill boomed or faded, Steptoe City followed suit. The town formed in the early 1890s and was named for Colonel E. J. Steptoe, an Indian fighter who rode with Beckwith. A post office opened in October 1893, with William Campbell serving as postmaster. Within a year, a Miners' Union Hall, a school, saloons, dance halls, and a red-light district had been established. Most of Steptoe City's residents were living here until a vacancy in either McGill or Ely opened up. The town reached its peak in the 1920s when its population topped 150. A "White Slavery" scandal hit Steptoe City and another nearby camp, Ragsville when it was discovered that some local saloons were operating a white slavery ring. This scandal contributed to the town's rapid decline. A fire in

1926 destroyed most of the town and sealed Steptoe City's fate. Some people remained for many years, however, and it wasn't until October 15, 1940, that the post office finally closed. Nothing remains at Steptoe City. The townsite is now covered by the McGill Smelter's tailing piles.

Stockville

DIRECTIONS: From Preston, take Nevada 318 north for 6.9 miles. Exit left onto U.S. 6 and follow for 3.4 miles to Stockville.

Stockville was originally one of many C. W. Mathewson ranches, established in the early 1870s. A post office opened on August 27, 1896, with George Hayden as postmaster, and Stockville became a shipping point for the White River ranches. The post office closed on January 7, 1899, but Stockville remained a prominent freight shipping point. During the early 1900s, a stage from Stockville to Hamilton ran twice a week. Few travelers took the stage, which acted more as a produce line from the ranches to the hungry miners in the Hamilton District. After the Hamilton boom collapsed, Stockville lost its importance. A small ranch continued to be active at the site. In the late 1940s, a gas station and small motel opened at Stockville and was the site's sole means of business. When the complex folded in the early 1970s, Stockville was abandoned for good. Today the station ruins and a couple of smaller buildings remain.

Strawberry

DIRECTIONS: From Cold Creek, head south for 9 miles. Exit right and follow for 1 mile to Strawberry.

Strawberry is a small ranching settlement that was first established by William Smith shortly after the White Pine boom. Smith had seven children, and the Strawberry school district was set up to educate his and other Newark Valley children. The Newark Valley voting precinct was also located here. The pride of Strawberry's owners, however, was the 12,000-tree orchard and huge fields of strawberries. The produce was shipped by four-horse wagons to Eureka and Hamilton. People picnicking high up on Newark Mountain not only could see the vast fields of red strawberries but also were treated to the beautiful scent of millions of strawberry flowers. A post office was active at Strawberry from 1899 to 1938. The original post office burned in 1917, and the office was moved into the ranch house. Today the present owners still have the original Strawberry Post Office sign in their possession. The ranch is still active, and original buildings, including a large stone barn with juniper beams of immense proportions, remain.

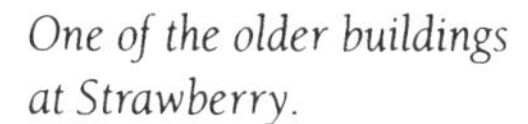
One of the older buildings at Strawberry.

Swansea

DIRECTIONS: *Located 1.5 miles above Shermantown.*

The prospecting frenzy that hit the Hamilton District in 1868 had men scouring the nearby mountains and canyons. Initial discoveries at Swansea, named for Swansea, Wales, were made in November 1868. In January Colonel Raymond and Judge Walsh laid out the Swansea townsite. Swansea boomed quickly, and the camp's residents hoped that it would become the smelting and milling center of the White Pine District. Despite the cold weather, Swansea continued to grow. In February Swansea's first smelter fired up but was unsuccessful. By spring Swansea had a population of 150 and two new smelters: J. J. Bassey's (started in May), and Abe Savage and Jeff Sherbourn's (started in April, burned on August 19). The summer months brought close to 300 people to the booming camp. The 10-stamp Vernon Mill was built by Lucas and Logan and began production on July 22. The mill cost $20,000 and was equipped with a Howland rotary crusher that processed 10 tons a day. The Swansea Mill, built by Perkins, Flint, and Company of San Francisco, was also constructed during the summer of 1869. The mill, with a capacity of 20 tons per day, contained two batteries of 5 stamps, eight receiving tanks, seven Varney pans, and two settlers. The mill processed ore from the Aurora Consolidated Mine on Treasure Hill.

By fall, Swansea had a population of 500, a small brewery run by "Swansea" Bill, and two new smelters, Massheimer and Company (White Pine Smelting Works) and the Jordan Smelting Works (started August 2). Swansea had a rivalry with nearby Shermantown. The camp tried to become prominent on its own, but most business establishments opened at better sites in Shermantown. Swansea peaked during 1870, and prominence for the district quickly shifted to Shermantown. Eventually Shermantown expanded and virtually annexed the dying camp of Swansea. Swansea became primarily a milling and smelting district, with most of the workers living in Shermantown. A new mill, the 40-ton Trinity, was built in 1872. The mill employed 10 but was forced to close during the winter months because the tailing piles froze. With the mining slowdown of the early 1870s, the smelters and mills in Swansea closed down one by one. By 1875 the camp had been completely abandoned. None of the mills ever reopened, although an occasional prospector lived in the abandoned stone houses. Today stone ruins and mill and smelter foundations mark the site.

Taft

DIRECTIONS: *From Cleveland Ranch, head north for 8 miles to Taft.*

Taft was one of the many Spring Valley post offices located at various ranches during a fifty-year period. As one ranch achieved prominence, the local post office was moved there, to stay until another ranch was deemed more worthy or the host ranch grew tired of the responsibility. The Taft post office opened on February 21, 1909, with John Yelland as postmaster, and closed on April 19, 1917. The school for Spring Valley was located here and still stands. It was open until recent times but has now outlived its usefulness.

Tamberlane Canyon (Nevada District)

DIRECTIONS: *From Ely, take U.S. 50 east for 7.7 miles. Exit left and follow for 2 miles. Exit sharp left and follow very rough road for 1.5 miles to Tamberlane.*

Joseph "Tamberlane Joe" Thompson made initial discoveries on Wagner Hill on April 25, 1869. Thompson and a few other prospectors were the only residents of the canyon until 1873, when a new strike brought about 40 men to the district. A small townsite was laid out and a camp began to form. A couple of frame buildings, including a saloon and a store, were built. Ore from Tamberlane's mines was sent to the Robinson Canyon 10-stamp mill. The richest mine was the Champion, discovered in 1880. Thompson was quite active in the district's mining affairs, locating and relocating mining claims. He left the district in 1887 and sold his timber rights to B. F. Miller, who cut and delivered wood to the Argus Mill in Taylor. Virtually all mining activity ceased in the late 1890s. Only occasional leaseholders worked the district until 1907 when Ceasar Cavigilia discovered manganese. Production from 1907 to 1930 was a shade under $200,000. The district was silent for close to twenty years until a mill was built here in 1951. Forty-five men were employed, and fairly consistent production took place until 1959, when the district was abandoned for good. Tamberlane Canyon had a total production of close to $400,000. Not much remains in the canyon now except some of the mill ruins.

Taylor

DIRECTIONS: From Ely, take U.S. 50 east for 13.5 miles. Exit left and follow for 3.5 miles to Taylor.

The earliest discoveries in Taylor Canyon were made by two men named Taylor and Pratt in 1872. The two were led to the Argus Ledge, soon to become the Taylor Mine, by Indian Jim Ragsdale, who was paid $500. The mine was sold to the Martin White Company of Ward in 1875 for $14,000. Another big discovery was the Monitor Mine, located by Robert Briggs, William Neil McGill, and W. G. Lyons. However, despite this activity, it wasn't until summer 1880 that Taylor finally began to boom. Henry and Fred Hilp opened a store, and Fanny Yates, of Lane City and Ward, moved here and built a new hotel. She was well liked because she grubstaked many prospectors in the district.

In 1881 Briggs (then a state senator from White Pine County), Lyons, and McGill organized the Monitor Mill and Mining Company and began construction on a 10-stamp mill near Steptoe Creek. Production at the Monitor Mine was stepped up to help pay for the mill. The stamps for the mill came from Seligman, near Hamilton, and the frame was shipped from Ward. The mill, equipped with a Knight water wheel, was started on September 12. Taylor really came into its own in 1883, with several new discoveries and more than $260,000 worth of ore shipped. A new company, the Argus, was formed, and construction on a new 15-stamp mill, located on Willow Creek, was begun. Joseph Caruthers, a mining and milling expert with the Argus Company, traveled east to get the Aultman Estate of Canton, Ohio, to advance the money for the Argus Mill. The main mines of the Argus Company included the Argus, Sunrise, Self-Cocker, and June. While the Argus Mine didn't have the richest ore, it did last the longest.

Part of the work crew at the Monitor Mine. (Bancroft Library)

Taylor reached its peak during 1883. The population increased to 1,500, and business included three general stores, four restaurants, three boarding-houses, a drugstore, and a doctor. A post office opened on May 9, with William Lyons as postmaster, and operated until September 9, 1893. In addition, by the end of the year, a brewery, an opera house, and a school had been built. Despite its size, Taylor was a peaceful town, with very little trouble. The town had two newspapers during its peak, the well-traveled *White Pine News* and the *Taylor* (formerly *Ward*) *Reflex*.

Taylor's free milling silver ore began to run out by 1885, and the decline accelerated during 1886. A minor disaster in the Argus Mine, when one miner was killed and another blinded after a drill hit some unexploded dynamite, hastened the decline. By the fall, Taylor's three main streets (Main, Argus, and June) were quickly emptying. The *Taylor Reflex* also left. The Monitor Mining and Milling Company sold its holdings in January to an English company, the Eberhardt-Monitor Mining Company, for $175,000. Eberhardt-Monitor had purchased the Argus Mine for $300,000 in 1884. The Argus Mill made one final shipment of $36,000 before closing on August 27, 1886. The mill was operated periodically until it was closed for good in 1889. The Monitor Mill was kept in production until fall 1886 but was dismantled when the company sold its holdings.

By 1888 only a handful of residents were left and most businesses and buildings had already moved to the new and booming town of Ely. By 1890 the only producing mines were the Argus and the Monitor. It wasn't until the Wyoming Mining and Milling Company entered the district in 1918 that Taylor finally revived. The company owned ten claims, including the Argus Mine, and in 1919 it constructed a 100-ton cyanide plant, started in April. During its first year the mill processed 60,000 tons of ore. The mill closed in early 1920 but was reactivated in August 1921 and treated 40 to 50 tons of ore daily for the next two years. Even after the Wyoming company left, leasing activity continued up until 1961. A mill was built near a new mine, the Enterprise, in 1939. The mill was refurbished in 1959 to treat ore from the Merrimac open pit.

Today Taylor is the site of a huge silver operation. The Silver King Mines, Inc., of Salt Lake City came into the district during the summer of 1980. A $10 million mine, mill, and processing facility was dedicated on June 19, 1981. An open-pit operation, it produces 1,200 tons of ore daily. Each month 80,000 ounces of pure silver are being produced. The owner of the company, Kaye L. Stoker, bought three square miles of claims in 1970 and estimated reserves should last for 25 years. The mine employs about 175 people, and a couple of businesses have opened at the town. Only two buildings remain from old Taylor. Most of the townsite has been taken over by the new operations. The ruins of the 1919 Argus Mill lie just north of the camp. Along Steptoe Creek are the Monitor Mill ruins, and at Willow Creek are remains of the Argus Mill. However, there is not much of interest at the townsite, and access to it is restricted.

Tippett

DIRECTIONS: From Stone House, continue on Nevada 2 for 17.4 miles to Tippett.

Tippett was founded by John Tippett during the early 1880s as a sheep ranch. Tippett, from Cornwall, England, was active in White Pine County's mining exploits. He owned claims in Glencoe and other nearby mining camps. Tippett married the sister of his mining partner and then settled down at his ranch. A biweekly stage between Cherry Creek and Tippett brought mail to the ranch. A post office, with Tippett as postmaster, opened on May 11, 1896, and served isolated ranches of Antelope Valley until June 30, 1926. Tippett died in the mid-1890s, and his widow married Gus Sellas. The couple organized the Tippett Land and Livestock Company. Though there have been many owners during the subsequent years, Tippett is still a ranching operation. Today, thoroughbred Arabian horses are the "livestock."

The small town of Tippett, with most of the town's population. (Nevada Historical Society)

The huge stone horse barn at Tippett. Horses for the Overland Stage were allegedly housed here.

The present owners are working hard to restore the ranch, which deteriorated during the 1960s and 1970s. While the ranch house is more recent, the highlight of Tippett is the old stone horse barn, built in the 1860s, long before Tippett was organized. The barn and small storehouse were used by both the Pony Express and the Overland Stage to house horses, which were switched at nearby Antelope Springs. The barn is a beautiful remnant of vintage Nevada history. Because of this fine structure and the enthusiasm of the present owners, Tippett ranks high on this author's list of White Pine's most enjoyable sites.

Treasure City (Treasure Hill) (Tesora)

DIRECTIONS: Located 1.5 miles below Hamilton.

The Treasure Hill District was the heart of the White Pine "excitement" during the late 1860s. Besides Treasure City, the Hill was host to many other small and short-lived mining camps, such as Picotillo, but these faded very quickly, and only Treasure City achieved any prominence. Initial discoveries were made by Leathers (blacksmith for the Monte Cristo Mining Company), Murphy, and Marchard on January 3, 1867, although the first discoveries in the White Pine District were made a few months earlier. This discovery, the Hidden Treasure Mine, was later sold to G. E. Roberts and Company for $200,000 in January 1870. Within one year of the Hidden Treasure discovery, 6,000 people were shivering in Treasure City. Another prominent mine in the district was the Keystone, discovered by John Turner on May 11, 1868. A reporter from the *White Pine News* visited the mine in August 1869 and said that the fantastic ore chamber had $8 million of ore in sight. The pure chloride silver assayed as high as $15,000 per ton.

Artist's view of Treasure Hill. (California State Library, Sacramento)

The town was originally named Tesora, and a post office with that name opened on April 23, 1869. It wasn't until June 15 that the name was officially changed to Treasure City. The post office was located in the basement of the Hallock and Meyers Store until a post office building was constructed in 1870. The first of many saloons in the town was the Aurora, run by Raines and Scott. Treasure City's tough terrain didn't slow its development at all. Most of the buildings were made of stone, not only to prevent fire but also to keep out the intense cold. The climate on top of Treasure Hill was extremely cold and bitter, but the lure of silver riches outweighed the unfavorable weather. Because of early milling problems, most ore was sent to Monte Cristo, Newark, or the Manhattan Mill in Austin. By the end of 1869, nearly two hundred mines and 10 mills, with 120 stamps, were active in the Treasure Hill District. At the peak of activity, 23 mills were crushing ore here. Water was brought in by the White Pine Water Company, beginning on October 9, 1869. There were two reservoirs on Treasure Hill, with a capacity of 60,000 gallons. However, the company shut off the water in March 1870 after reportedly losing $4,000 a month.

Most mines besides the Hidden Treasure and Keystone mines were located farther down the hill. But there were many producing mines near Treasure City, including the Argyle, Pocotillo, Blue Belle, Virginia, Emersley, and California. The mining excitement in the district helped propel Treasure City to its peak during 1869 and 1870. The town's main street stretched for almost

One of the many spectacular stone ruins at Treasure City.

This stone cabin commanded a beautiful view.

three-quarters of a mile. Its many stores were crowded along the crest of the hill, with mining claims and small shafts located only a few feet away. The top of Treasure Hill was almost completely barren, and cold winds constantly whistled through the town. During the hustle and bustle of 1869, Treasure City had more than 40 stores, about a dozen saloons, impressive Masonic and Odd Fellows lodges, and its own stock exchange. All of this excitement, however, did not save the mines. The ore values plummeted and many of the deposits disappeared. By the end of 1870 Treasure City's population had shrunk to fewer than 500 people. Several businesses remained, living off patrons from the nearby towns of Hamilton, Shermantown, and Eberhardt. A major fire in spring 1874 leveled most of Treasure City and destroyed the business district, and because of the depression that settled on the city, very little was rebuilt. By 1880, only 24 people were living here. The post office closed on December 9, 1880, and during the following years only a handful of residents weathered the winters in Treasure City. Total production from the Treasure Hill mines from 1867 to 1880 was close to $20 million.

There was a small revival in the early 1920s when the Treasure Hill Deposits Mines Company, with N. A. Dunvon as president, acquired 125 acres of claims and sank a 250-foot shaft. During its first two years, the company produced $1.5 million, but production dropped quickly and the company folded in 1927. Today Treasure Hill is once again the scene of intense mining activity. Extensive cyanide leaching operations have made the hill echo with the sound of huge ore trucks and bulldozers. While the remains of Treasure City are absolutely spectacular, visitors must be extremely careful about the flow of traffic lest they be heading up the hill when a 20-ton ore truck is coming down! Because of its altitude, Treasure City presents some stunning photographic opportunities. Besides numerous stone structures and mine ruins, the site offers a breathtaking view of White Pine County and the Treasure Hill Mining District. Treasure City, and all the district, is a must for the visitor's itinerary. Treasure City and the other nearby ghost towns (Hamilton, Seligman, Shermantown, Eberhardt, Swansea, and Monte Cristo) merit a visit of at least two days to thoroughly explore their interesting features.

Tungsten Mines (Hub) (Lincoln)

DIRECTIONS: *From Shoshone, head north for 7 miles. Exit right and follow for 3 miles to Tungsten Mines.*

Original discoveries were made by W. H. Buntin in 1869, but production was poor. The district was pretty much idle until 1900, when a mining district was organized. In 1904 the Tungsten Mining and Milling Company purchased all claims. In 1909 the company was bought out by the Heubreite-Tungsten Mining Company, renamed the U. S. Tungsten Corporation in 1910. A small camp of 50 sprang up during 1910 and a 50-ton concentrating mill was built. Plunging ore prices forced closure of the mill in 1911, but it was restarted in 1915. A 6,000-foot ditch from Williams Creek was dug to supply the mill with water. A post office opened on October 14, 1916, with Ezra Ramsay, Jr., as postmaster, but ore production fell drastically by the end of that year. The post office closed on June 30, 1917, and the mill stopped operations about a month later. The U. S. Tungsten Corporation folded after producing $704,000, and the camp was completely abandoned by fall 1917. Any hopes of a revival were crushed when the mill was dismantled in the early 1920s. Today a couple of old shacks, mill foundations, and tailing piles mark the site.

Tungstonia (Kern) (Eagle)

DIRECTIONS: *Located 1.5 miles southwest of Parker.*

Tungstonia came into existence after George Sims and C. Olsen discovered tungsten in 1910. It wasn't until 1914, however, that steady mining production began. Two Utah-based companies, the Salt Lake Tungstonia Mines Company and the Utah-Nevada Mining and Milling Company, became

The Utah-Nevada Mining and Milling Company mill at Tungstonia, 1917. (USGS)

active in the area, and both began to construct mills in 1915. The Salt Lake company, with L. W. Roberts as president, controlled seven claims adjoining the Shepherd, or Ophir, Mine. The company built a 20-ton mill, which consisted of a Marcy ball mill and Wilfey tables. The mill was enlarged to 25 tons in 1916 and was in service around the clock. The concentrates contained 75 percent tungstic oxide and were shipped on the Deep Creek Railroad. The main development for the company was a 300-foot tunnel that had ten fissure veins. Utah-Nevada Mining and Milling began construction of its mill in 1915, but the company's financial backing fell apart and the mill was never completed.

The camp of Tungstonia grew to about 25 during 1915, and by the summer of 1916, the camp had a population of 50, a couple of saloons and stores, a boardinghouse, and several wood-frame buildings. A post office, with Kirby Smith as postmaster, opened on January 4, 1917, but closed on August 3. The Salt Lake company left the district in early 1918 and most of Tungstonia's residents moved. The Griffin Mining Company leased the property during the summer but produced no ore. Tungstonia retained only a few lonely prospectors once the mining activity ceased. The district was abandoned until 1935 when new tungsten operations were begun. The new mines were the Dandy, the Whiskey Bottle, and the Tungstonia. In 1938, a 36-ton mill was built, which was active until 1942. After 1942 only occasional leasing activity took place. Total production from the Tungstonia district was $126,000. Today ruins of the three mills and some scattered rubble mark the site.

Uvada (Pleasant Valley) (Parker)

DIRECTIONS: From Parker, continue north for 2 miles to Uvada.

Pleasant Valley was the site of many ranches beginning in the 1890s. A post office, named Pleasant Valley, was opened at one of the ranches on March 15, 1892, with William Mallany as postmaster. The ranch, as were most in the valley, was run by the Henriod brothers. The few small mines in the district included the Mint (owned by John Tippett and William Mallany), the Tiger and Blackjack claims (run by the Henriod brothers), and the Blue Mass Mine. In June 1910 a white man working at the Blue Mass was badly beaten by an Indian, and he retaliated by shooting his attacker. Fears of reprisals never materialized, however, and tensions were finally defused. The post office, which had closed on April 26, 1894, reopened at Parker, another Pleasant Valley ranch, on January 13, 1910, with Agnes Henriod as postmistress, and operated until April 1, 1927. On May 9, 1928, the post office was once again reopened, this time at nearby Uvada, and remained open until June 30, 1944. Today the range in Pleasant Valley is still used, but most of the ranches are abandoned. These old ranches are scattered throughout the valley, and vintage buildings remain.

Veteran

DIRECTIONS: Was located 0.5 mile west of Ruth.

In 1906 copper ore was discovered, and soon a tent city, complete with bunkhouses and a saloon named the Veteran, grew up. Veteran was the company town of the Cumberland-Ely Copper Company. The Veteran Mine was the main producer for the company but never really lived up to the potential the owners thought the mine had. Veteran was the western terminus of the Nevada Northern Railway, and a large depot was constructed. A three-story mining office was also built, as well as two-story wooden bunkhouses for the 100-plus miners who were employed in the six-compartment Veteran Mine. Ore values fell during 1913, and the mine closed a year later. Soon after that, all of Veteran's buildings were moved to Ruth, and work began on the huge Veteran open-pit copper mine. The Veteran Pit is discussed further in the section on Ruth's history. Now only that huge pit is left to mark the camp of Veteran.

The Veteran Mine. (U.S. National Archives)

Ward

DIRECTIONS: From Ely, take U.S. 50 east for 5.6 miles. Exit right and follow for 7 miles. Exit right and follow for 3 miles to Ward.

Two teamsters, William Ballinger and John Henry, working for the Toano-Pioche stage made initial discoveries in March 1872. Their finds were developed into the Paymaster Mine, which soon made Ward the fastest-growing camp in eastern Nevada. The camp was named after B. F. Ward, who, along with George Tyler and Ben Mitten, all from Mineral City, located the Ward townsite. By 1875 Ward was the largest town in White Pine County, boasting a population of 1,000. Fanny Yates, of Lane City, moved here and built a large, fancy hotel. The first merchants in Ward were the famous Nevadan grocers Henry and Fred Hilp. Education was a priority here, and in 1875 an abandoned red-light-district house was converted to a schoolhouse. All mines were purchased by the Martin White Company of San Francisco during the summer of 1875, and that company was the main producer for many years.

Another big mine, the Pleiades, was discovered by Jack Roach and Mat Gleason, both of Mineral City. Soon after the discovery, however, the two became embroiled in a bitter dispute over percentages, which ended when Roach killed Gleason. The mine was then purchased by the Martin White Company. During 1876 Ward continued to boom. A 20-stamp mill was

The town of Ward during its peak. (Nevada Historical Society)

moved from Troy (Nye County) at a cost of $85,000. The mill, however, was a failure. Another $35,000 was sunk into the project before it was finally abandoned. Two newspapers, the *Ward Miner* and the *Ward Reflex,* began publication. Discoveries at Ward helped propel the town to its peak during 1877.

Ward's population reached a high of close to 2,000. A post office opened on January 2 and was moved five times, three because of fires, before it closed on September 7, 1887. Also in 1877 a city hall was constructed and Wells-Fargo opened an office. The town was kept virtually crime-free by the 601 Vigilantes. The name came from six feet under, no trial, one rope. The 601 meted out quick justice, and Ward's crime rate dropped to zero.

Besides the Martin White Company, other companies that moved into the district during 1877 included the Steptoe Consolidated Mining Company (Emily, Profit, Ready Cash, Light, Robert Briggs, Cow, Old Ned, and Fourth of July mines), Weaver, McKinzie, and Company (Pleiades and Atlantic Cable mines), Governor Consolidated Mining Company (Governor and Grampus mines), and Ward Consolidated Mining Company (Shark, San Mateo, and IXL mines). The mines run by the Martin White Company included the Paymaster, Defiance, Mountain Pride, Carolina, Young America, Mammoth, Home Ticket, Wisconsin, Maggie, Cyclops, and Alameda. The Paymaster was the big mine in 1877, producing $194,000.

Conflict concerning mine claims came to a head in February when fourteen men at the Ward Consolidated Mine were driven off by armed men from the Martin White Company. However, the Ward Consolidated men were able to retake the mine and hold it until the matter was resolved, in favor of the Ward company. Ward began to decline as 1878 progressed. Nevertheless, the Martin White Mill began production on November 16, 1878. The mill was actually converted to 20 stamps from a smelter because the lead content of the ore fell dramatically as the ore deposits played out. Equipped with three White and Howell furnaces, the mill operated only until fall 1879. It was started up again for a short while during the summer of 1883.

The combination of vanishing ore deposits and a new boom at Cherry Creek spelled doom for Ward. By 1880 the population had shrunk to 250. A huge fire on August 18, 1883, destroyed the city hall, the school, and virtually all of downtown Ward. A week later the *Ward Reflex* moved to the boomtown of Taylor, striking another blow against Ward's survival. By 1885 only one business was left in town, and the population stood at 25. Ward was basically a dead town until 1906, when all Martin White holdings were sold to the Nevada United Mines Company. Besides 12,000 acres of mining ground, the company also bought 1,000 acres of ranchland at Willow Creek, near the famous Ward Charcoal Kilns. Over the next six years, Nevada United spent $500,000 in developing the Ward mining interests. Active mines included the Paymaster, Ready Cash, Defiance, Mammoth, Mountain Pride, Caroline, Young America, and Shark. While production was not spectacular, it was consistent for years. The best year was 1917, with 2,800 tons of ore produced each month. Shipments from the mines continued until September 1920. During this activity, the population stood at around 50, but as soon as operations were curtailed, the town emptied. A couple of brief revivals occurred in the 1930s and 1960s, but little production was recorded. Today Ward is an active mining site, operated by the same company that has reopened mines at Taylor and Hamilton. Unfortunately the town is now off limits, fenced off by the company. A couple of buildings, smelter ruins, and mill foundations remain. The Ward Cemetery, one mile east of the townsite, is well worth a stop. Many interesting wooden markers remain and are partially legible. It is very sad to note that many of the graves are those of very young children who fell victim to the many different diseases prevalent in early mining camps. The fabulous Ward Charcoal Kilns, now a Nevada state park three miles south of Ward, are a must for the visitor.

Yelland

DIRECTIONS: From Cleveland Ranch, head north for 6 miles to Yelland.

Yelland was founded by John "Josh" Yelland, a sheepherder from Cornwall, England, in 1908. Yelland came to Cherry Creek when he was twenty, and in 1889 he purchased the Nigger Creek Ranch in Spring Valley. In 1908 he sold out and took over the Taft Creek Ranch, which became Yelland. John Yelland had two boys and two girls. His son Henry James was the first Nevadan killed in World War I. A post office opened on November 17, 1924, and served the ranches of Spring Valley until closing on January 15, 1927. The ranch is still in the Yelland family and is presently run by Marian Yelland, whose late husband, Louis, was the son of John. Several of Yelland's original buildings remain at the site.

BIBLIOGRAPHY

The books listed below represent only the major sources used in compiling this book. Countless manuscripts, files, and personal interviews also served as excellent references, but to list all of these other sources would be much too lengthy.

Newspapers

Battle Mountain Messenger (Battle Mountain)
Central Nevadan (Austin)
Eureka Daily Leader (Eureka)
Eureka Daily Republican (Eureka)
Eureka Sentinel (Eureka)
Lewis Weekly Herald (Lewis)
Mining and Scientific Press (San Francisco)
People's Advocate (Austin)
Reese River Reveille (Austin)
Ruby Hill Mining News (Ruby Hill)
White Pine News (Ely, Cherry Creek, Taylor)

Literary Sources

Abbe, Donald. *Austin and the Reese River Mining District.* University of Nevada Press, 1985.

Adair, Donald. *Geology of the Cherry Creek District, Nevada.* Privately published, 1961.

Albert, Herman. *Odyssey of a Desert Prospector.* University of Oklahoma Press, 1947.

Angel, Myron. *Thompson & West's History of Nevada, 1881.* Howell-North Books, 1958.

Ashbaugh, Don. *Nevada's Turbulent Yesterday.* Westernlore Press, 1963.

Baily, Edgar H., and David A. Phoenix. *Quicksilver Deposits in Nevada.* University of Nevada, Reno Bulletin, no. 41, 1944.

Bancroft, Hubert H. *History of Nevada, Colorado, and Wyoming, 1540–1888*. The History Company, 1890.
Blatchly, M. E. *Mining and Milling in the Reese River Region*. Slate and Jones, Stationers, 1867.
Bloss, Roy. *Pony Express—The Great Gamble*. Howell-North Books, 1959.
Couch, Bertrand, and Jay Carpenter. *Nevada's Metal and Mineral Production (1859–1940)*. University of Nevada Bulletin, vol. 37, no. 4, 1941.
Davis, Samuel P. *The History of Nevada*. Elms Publishing Company, 1913.
Egan, Howard. *Pioneering the West, 1846–1878*. Howard Egan Estate, 1917.
Elliott, Russell. *The Early History of White Pine County, Nevada, 1861–1887*. White Pine District Library Association, 1965.
Emmons, W. H. *A Reconnaissance of Some Mining Camps in Elko, Lander, and Eureka Counties, Nevada*. USGS Bulletin, no. 408, 1910.
Ferguson, Henry. *The Mining Districts of Nevada*. Nevada Bureau of Mines Bulletin, no. 40, 1944.
Fink, Kendrick. *Geology and Ore Deposits of the New Pass Mines, Lander County, Nevada*. Privately published, 1976.
Fleming, Jack. *Copper Times: An Animated Chronicle of White Pine, Nevada*. Jack Fleming Publishers, 1987.
Florin, Lambert. *Ghost Towns of the West*. Superior Publishing Company, 1971.
Fox, Theron. *Nevada Treasure Hunters Ghost Town Guide*. Harlan-Young Press, 1961.
Frank, Ralph. *Cherry Creek Reminiscence*. Privately published.
Frickstad, Walter, and Edward Thrall. *A Century of Nevada Post Offices*. Pacific Rotoprinting Company, 1958.
Gallagher, Charles. *Memoir and Autobiography*. Oral History Department, University of Nevada, Reno, 1965.
Gianella, Vincent. *Bibliography of Geologic Literature of Nevada*. University of Nevada Bulletin, 89:6, 1945.
Group of Valuable Silver Mines at Austin, Reese River, Lander County, Nevada. George Rand, Printers, 1865.
Harris, Robert. *Nevada Postal History*. Bonanza Press, 1973.
Heidrick, Tom. *Geology and Ore Deposits of the Ward Mining District*. Privately published, 1965.
Hill, J. M. *Notes on Some Mining Districts in Eastern Nevada*. USGS Bulletin, no. 648, 1916.
Jackson, W. Turrentine. *Treasure Hill*. University of Arizona Press, 1963.
Kelly, J. Wells. *First Directory of Nevada Territory (1862)*. Talisman Press, 1962.
King, Buster. *The History of Lander County*. University of Nevada, master's thesis, 1954.
Knudtsen, Molly. *Here Is Our Valley*. University of Nevada, 1975.
———. *Under the Mountain*. University of Nevada Press, 1982.
LeMaire, Rene. *Recollections of Life in Lander County, Nevada*. Oral History Project, University of Nevada, Reno, 1970.
Lewis, Oscar. *The Town That Died Laughing*. Little, Brown, 1955.
Lincoln, Francis Church. *Mining Districts and Mineral Resources of Nevada*. Nevada Newsletter Publishing Company, 1923.
Lingenfelter, Richard. *The Newspapers of Nevada*. John Howell Books, 1964.
Mack, Effie Mona. *Nevada*. Arthur H. Clark, 1936.
Mason, Dorothy. *The Pony Express in Nevada*. Harrah's, 1976.

Metal and Non-Metal Occurrences in Nevada. University of Nevada Bulletin, vol. 26, no. 6, 1932.
Molinelli, Lambert. *Eureka and Its Resources*. Lambert Molinelli and Company, 1879.
Murbarger, Nell. *Ghosts of the Glory Trail*. Desert Magazine Press, 1956.
Myrick, David. *Railroads of Nevada, Vol. 1*. Howell-North, 1963.
Paher, Stanley. *Nevada Ghost Towns and Mining Camps*. Nevada Publications, 1973.
Raymond, R. W. *Report on the Mineral Resources of the States and Territories West of the Rocky Mountains*. U.S. Government Printing Office, 1875.
Read, Effie O. *White Pine Lang Syne*. Big Mountain Press, 1965.
Reichman, Frederick. *Early History of Eureka County*. University of Nevada, master's thesis, 1954.
Roberts, Montgomery. *Geology and Mineral Resources of Eureka County, Nevada*. Nevada Bureau of Mines Bulletin, no. 64, 1967.
Schrader, F. C. *Cherry Creek District, White Pine County*. University of Nevada Bulletin, vol. 25, no. 7, 1931.
Smith, Roscoe. *Mineral Resources of White Pine County, Nevada*. Nevada Bureau of Mines Bulletin, no. 85, 1976.
Stager, Harold. *Geology and Mineral Deposits of Lander County, Nevada*. Nevada Bureau of Mines Bulletin, no. 88, 1977.
State Mineralogist of Nevada. *Annual Reports, 1864–1928*.
Stevens, Horace. *The Copper Handbook: A Manual of the Copper Industry of the United States*. Vols. 2–10. U.S. Government Printing Office, 1908.
Vanderburg, W. O. *Placer Mining in Nevada*. University of Nevada Bulletin, vol. 30, no. 4, 1936.
Vanderburg, W. O., and Alfred Smith. *Placer Mining in Nevada*. University of Nevada Bulletin, vol. 26, no. 8, 1932.
Vander Meer, Cornell. *Vanishing Frontiers, White Pine County*. Special Collections, Library, University of Nevada, 1977.
Van Nostrand, D. *The Silver Mines of Nevada*. Wm. C. Bryant and Company, 1865.
Weed, Walter. *The Copper Handbook: A Manual of the Copper Mining Industry of the World*. Vol. 2. U.S. Government Printing Office, 1912.
———. *The Mines Handbook and Copper Handbook*. U.S. Government Printing Office, 1916–26.

INDEX